This book provides a clear and concise account of the physiology and form of the fish circulatory system. Emphasis is primarily placed on the function of the system although details of structure have been included. Following some revisionary ideas on haemodynamics, attention is focussed on the heart as the primary pump in the fish circulatory system. The fine structure and the electrical and ionic changes in the cardiac myocytes are described and the major events of the cardiac cycle are outlined. This is followed by a description of the structure of the peripheral vessels and of circulation in certain special areas such as the gills, renal portal system and the secondary blood system. Further chapters are devoted to the blood and the haemopoietic tissues and include an account of the different types of retial system that concentrate oxygen or heat in various parts of the body.

This book is up-do-date, well illustrated and written in a style comprehensible to anyone with a basic knowledge of the biological and physical sciences. Both undergraduate and graduate students of physiology, zoology and marine science will find this an invaluable reference text.

Physiology and form of fish circulation

PHYSIOLOGY AND FORM OF FISH CIRCULATION

Geoffrey H. Satchell

Department of Physiology, Otago Medical School, New Zealand

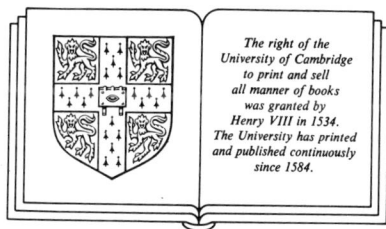

The right of the
University of Cambridge
to print and sell
all manner of books
was granted by
Henry VIII in 1534.
The University has printed
and published continuously
since 1584.

CAMBRIDGE UNIVERSITY PRESS

Cambridge   New York   Port Chester

Melbourne   Sydney

Published by the Press Syndicate of the University of Cambridge
The Pitt Building, Trumpington Street, Cambridge CB2 1RP
40 West 20th Street, New York, NY 10011-4211, USA
10 Stamford Road, Oakleigh, Melbourne 3166, Australia

First published 1991

Printed in Great Britain at the University Press, Cambridge

British Library cataloguing in publication data
Satchell, Geoffrey H.
Physiology and form of fish circulation.
1. Fish. Circulatory system. Physiology
I. Title
597.011

Library of Congress cataloguing in publication data
Physiology and form of fish circulation /
Geoffrey H. Satchell.
p.   cm.
Includes bibliographical references and index.
ISBN 0 521 39519 4 (hardback)
1. Fishes – Cardiovascular system.
2. Blood – Circulation.
I. Title.
QL639.1.s28   1991
597'.011 – dc20   90-38675   CIP

ISBN 0 521 39519 4

CE

This book is dedicated to
PROFESSOR PATRICK J. MOLLOY
Professor of Cardiac Surgery
The Otago Medical School

CONTENTS

Contents

In recent years mankind has become increasingly aware of the finite nature of our natural resources. At least a third of the world's population is insufficiently nourished. At the same time much of the world's land suitable for farming is already under cultivation and it seems unlikely that the ever increasing shortfall of animal protein can be produced on farms. The oceans produce only 1–2% of the calories consumed by man and world fisheries too are fast approaching the point where all the well-defined stocks of fish are fully utilized. Perhaps an annual sustainable yield of 100 million tons of fish is a possibility; it seems unlikely that this catch can be doubled. Indeed, the growth of sea fisheries declined between 1975 and 1980. Today we still get most of our fish by hunting them with baited hooks and nets. More than a thousand years ago mankind realised that hunting was an inefficient way of obtaining meat, compared with farming it and in many countries the potential for aquaculture, i.e. the controlled cultivation and harvest of fish, is under investigation.

Aquaculture assumes that the proper management of systems, in terms of inputs of high quality water and feed, will give higher yields than unmanaged natural systems. An essential requirement of such management is an understanding of the physiology of the organisms to be cultivated. Established texts exist concerning the physiology of most of man's domesticated mammals and much of the success in raising meat from them is due to this; those wishing to know about the physiology of fish are less well provided. This book has been written to provide a comprehensive review of the physiology of the fish circulatory system. It is intended to serve the needs of students in Fisheries Biology but it will also be useful for courses in Comparative Physiology. The interrelation of structure and function are nowhere more evident than in the circulatory system, and the text and illustrations include structural descriptions where these will help the student to grasp the physiology. This necessity is the greater in fish circulatory physiology because features such as the retia mirabilia which concentrate oxygen in the eye and swimbladder and the separate and partly independent secondary blood system, do not occur in higher vertebrates and are often unknown even to circulatory physiologists.

I have borne in mind that some readers, interested in general physiology, may be unfamiliar with the scientific names of the many species of fish that have been used in circulatory studies. For this reason I have used common names, and have included the latin name only when it is first mentioned in the text. The rainbow trout *Salmo gairdneri*, has been the species most frequently studied by physiologists and I have referred to it simply as 'the trout'; the brown trout *Salmo trutta*, and brook trout *Salvelinus fontinalis*, I have referred to as such. I have tried to spare my readers from the arcane scholarship of the fish systematists; I am aware that the rainbow trout has, by some authorities, been placed in the genus *Oncorhynchus*. I am also aware that in none of the 44 papers cited in the references, which include the scientific name of the rainbow trout in the title, is the name *Oncorhynchus* used. At the end of the book I have included an Appendix in which both common and scientific names are given.

This book would not have been possible without the help of many people. I am quite particularly indebted to my colleagues in the Physiology Department of the Otago Medical School who have been so generous with their time; I wish specifically to thank Dr E. R. Fawcett, Dr C. P. Bolter and Dr J. P. Leader for some valuable discussion. I am indebted to Jean Clough and Douglas Sanderson for technical help with the illustrations. Professor A. P. Farrell, of Simon Fraser University, Professor R. M. G. Wells of the University of Auckland, and Dr M. E. Forster of the University of Canterbury have read much of the manuscript and have been ever helpful with criticism. Thanks are also due to Professor H. A. Bern of Berkeley, Professor K. R. Olson of the Indiana University School of Medicine and Dr P. S. Davie of Massey University, who have helped me with specific points. None of these gentlemen is to be held responsible for what I have written, but to all of them I offer my sincere thanks for their advice and encouragement. Inevitably, in a book of this length, mistakes will have been made and I would be grateful to my readers if they will write to me and let me know of them.

G. H. Satchell
Department of Physiology
Otago Medical School
January 1990

1

INTRODUCTION

Fish are the largest class of vertebrates; with over 21 000 existing species they outnumber the amphibians, reptiles, birds and mammals put together. Fish span a long period of geological time for the earliest hagfish are now believed to have arisen in the Cambrian period, (Hardisty 1982). As inhabitants of both salt and fresh water they are spread over 71% of the earth's surface. The greatest depths of the oceans are deeper than the world's highest mountains, so fish span a greater vertical range than air-breathing vertebrates. For this reason they endure a larger range of pressures; on land the pressure at 9000 m is 0.3 atm but in the greatest depths of the oceans it is 1200 atm. They can live continuously in waters of extreme temperature. Some of the Chaenichthydae live permanently below the polar ice at temperatures of -1 to $-1.8\,°C$; a cichlid *Oreochromis grahami* lives in alkaline lakes in Kenya's rift valley, in which the water has a temperature of 43 °C and a pH of 9.6–10.5 (Johansen *et al.* 1975). Fish have become adapted to environments quite as extreme as any occupied by higher vertebrates.

Fish, like other vertebrates, possess a circulatory system consisting of a pump, the heart, and a continuous system of branching tubes, the arteries, arterioles, capillaries and veins, which form a closed circuit so that all of the blood leaving the heart is returned to it. The blood flowing through it carries oxygen from the gills to the tissues, carbon dioxide to the skin and gills to be eliminated, the soluble products of digestion from the gut to the liver, more processed nutrients from the liver to the tissues and ions such as Na^+ and Cl^- which are crucially involved in osmoregulation. It carries hormones and vitamins; certain specialized carrier proteins occur in the plasma, such as transferrin, which transports iron, and caeruloplasmin, which transports copper.

The primary role of the circulatory system is thus to maintain a flow of blood through the gills and tissues of the body, and to return it to the heart. The flow of blood through a system of tubes is governed by the same laws as operate in physical systems. The structure and physiology of any circulatory system reflects the primacy of these laws.

Some elementary haemodynamics

Flow, denoted by the symbol F, is measured in units such as ml min^{-1} and is determined by pressure P, and resistance R, in a manner analogous to Ohm's law.

$$F = \frac{\Delta P}{R}$$

In the equation above, pressure is shown as ΔP, to indicate that it is the difference of pressure along the length of the vessel that effects flow. In a short length of tubing this would be equal to $P_1 - P_2$, i.e. the pressure difference between the two ends of the tube. The absolute pressure to which the system is subjected is irrelevant. Fish in the ocean depths are at very great pressures but these affect tissues and blood vessels alike and do not result in a gradient of pressure along particular vessels.

A further distinction needs to be made between driving pressure and transmural pressure. Pascal's first law tells us that any one level pressure acts equally in all directions. Inside a blood vessel there will be a component acting along its length; this is the ΔP indicated above and provides the driving pressure that effects flow. A component at right angles to this exerts a pressure on the vessel wall and is opposed, to a greater or lesser extent, by a pressure exerted by the wall and the tissues that press against it. The difference between the two is the transmural pressure; if it is positive it will distend the vessel to an extent dictated by the elasticity of its wall. If it is negative, the vessel will collapse, unless, as happens in the caudal vein and some of the cutaneous veins, this is prevented by the rigidity of the wall and surrounding tissues.

Driving pressure and transmural pressure are really independent. Any restriction to the outflow from a vascular bed will raise the pressure in its veins, capillaries and arteries alike. The transmural pressure will rise accordingly but there may be little or no change in driving pressure. Increase or decrease in transmural pressure has the potential to dilate or contract vessels. Teleosts have the ability to constrict their efferent gill vessels in response to hypoxia; this raises the transmural pressure in the gill lamellae and, by dilating them, increases the area available for gas exchange (Nilsson and Pettersson 1981). Elasmobranch fish have a prominent sphincter surrounding the opening of each hepatic vein into the sinus venosus. Johansen and Hanson (1967) report that this is dilated by catecholamines, which (Chapter 10 and 11) are liberated into the blood of elasmobranchs during exercise and hypoxia. Dilating the hepatic vein sphincters will allow the transmural pressure in the large sinuses of the hepatic and hepatic portal veins to fall and blood to be transferred from the visceral to the somatic circulation.

The Poiseuille equation

Flow through a tube may be pictured as the movement in it of concentric laminae of fluid, with the central core moving most rapidly, and the outermost layer stationary against the wall. All the resistance to flow is caused by the resistance to sliding of the laminae of fluid, one upon another. It is not, at the velocities of flow likely to be present in the vessels of fish, due to friction against the wall of the vessel. Because narrow tubes have a smaller proportion of their volume in the faster-moving axial stream, resistance is greatly influenced by the radius of the tube, r, as well as by its length, L and by the viscosity of the fluid, η. Poisueille's equation tells us that

$$F = \frac{\Delta P \pi r^4}{8 L \eta}$$

Because flow is proportional to the fourth power of the radius, small changes in this cause large changes in flow. A reduction in the radius of a vessel by 16% will halve the flow through it. Poiseuille's equation strictly applies to flow through rigid unbranched tubes, and is thus not entirely appropriate to the fish blood system. The arteries that take blood to the gills, for example, are not rigid, but very compliant as they are rich in elastic tissue; moreover, they branch repeatedly.

Viscosity

Blood is a fluid and a fluid may be defined as a substance which cannot indefinitely withstand a force tending to deform it and to cause one layer of it to slide over another; i.e. a shearing force. This tendency to change shape is termed viscosity, η, and Sir Isaac Newton in the *Principia* of 1687 defined it, with perfect clarity, as a lack of slipperiness ('*defectus lubricitatis*') between adjacent layers of fluid. Resistance to shear depends on the rate of change of deformation, the shear rate, not on the actual distance moved. The shear rate is the gradient of velocity and is expressed in units per sec; it is derived from cm sec^{-1} cm^{-1}. The unit of η in the CGS system is the poise and is defined as the tangential force per unit area (dynes cm^{-2}) when the shear rate is unity i.e. 1 cm sec^{-1} cm^{-1} and, in the haematological literature, the viscosities of bloods are normally expressed as so many centipoises, (0.01 poise). I have therefore retained this unit, from the CGS system, rather than the SI unit of Pascal seconds (Pa sec); one centipoise = 1×10^{-3} Pa sec. Water at 20 °C has a η of approximately 1 centipoise, (1 cp): the bloods of fish have η values within the range of 10–90 cp. Fluids such as saline, in which η is independent of the shear rate, are termed Newtonian fluids. Blood has cells in it, and the large protein molecules in the plasma tend to attach to these and link them together when the blood is flowing slowly. Hence the η increases as the rate of shear decreases: it is an example of a non-Newtonian fluid. The η of blood also increases as its content of red cells is increased. As all of the resistance to flow of the blood is due to its viscosity, we can see that there may well be an optimal haematocrit, i.e. red cell content. Increasing the number of red cells will increase the capacity of the blood to transport oxygen, but will also increase the work the heart has to do to propel it through the circulatory system; indeed when the haematocrit exceeds 70%, blood can hardly be regarded as a fluid. We will return to this topic in Chapter 4.

Total fluid energy

Poiseuille's equation is an over-simplification in that it omits the fact that a column of moving blood has kinetic energy; at any particular point in a blood system, pressure energy is not the sole determinant of flow; the energy of the moving mass of blood needs also be to considered. We can include this by stating that flow depends on differences, along the length of a vessel, in the total fluid energy of a unit volume of blood (Hicks and Badeer 1989). The pressure component in this is commonly the largest, but in low pressure, e.g. venous, systems the kinetic component may be a considerable proportion of the total. The equation is

$$E = P + pgh + \tfrac{1}{2}pV^2$$

E = total energy in Joules cm^{-3}, P = pressure in mm Hg, V = velocity in cm sec^{-1}, p = density in g cm^{-3}, g = acceleration due to gravity = 980 cm sec^{-2}, h = height in cm above a datum level

The term pgh may be quite large in a land vertebrate, in which the height of the body above the ground may be considerable, but pgh is likely to be very small in a fish. This is because most fish have their greatest dimension, i.e. length, horizontal and because the gravitational pressure of the blood is largely cancelled out by the buoyancy supplied by the surrounding water. But it may be that during sudden accelerations and decelerations transients are developed in which the acceleration due to gravity is replaced briefly by forces generated by the movement of the fish. In the blue fish, *Pomatomus saltatrix*, the acceleration is up to $3G$ and Ogilvy and DuBois (1982) suggest that such forces could have led, in the course of evolution, to a toleration of the force of gravity.

Flow will occur along a blood vessel from A to B if the total energy, at the point A exceeds that at B, and in certain situations, such as the base of the mammalian aorta, flow may be against the gradient of pressure. The two terms, P and $\frac{1}{2}pV^2$ are to some extent interconvertible. The complex fourth chamber of the teleost heart, the bulbus arteriosus, (Chapter 2) is, it seems, a device for converting kinetic energy to pressure energy during systole and changing it back again during diastole. As we shall note shortly, the same volume of blood has to flow through each segment of the circulatory sytem but the proportions of the pressure and kinetic terms are widely different in the arteries and veins. In the central veins, where pressures are ambient or even subambient, kinetic energy must be a substantial part of the total, although we as yet lack the data to say what this fraction is.

Resistance

Resistance may be calculated by dividing pressure by flow and thus has arbitrary units of mm Hg ml^{-1} (of flow) min^{-1} kg^{-1} (of fish body weight). The greater the pressure required to propel 1 ml of blood through the vascular bed, the higher the resistance must be. Work done against resistance is ultimately dissipated as heat and is lost to the system. There are thus advantages in keeping the total peripheral resistance, TPR, as low as possible. When resistances are connected in series, the TPR is simply the sum of their separate values.

$$\text{TPR} = R_1 + R_2 + R_3 + \ldots R_n.$$

When resistances are connected in parallel the reciprocal of the TPR is the sum of the reciprocals of their separate values.

$$\frac{1}{\text{TPR}} = \frac{1}{R_1} + \frac{1}{R_2} + \frac{1}{R_3} + \ldots \frac{1}{R_n}$$

Clearly, the 'in parallel' arrangement results in a much lower TPR. Three vascular resistances each of 100 resistance units have, in series, a TPR of

300: in parallel they have a TPR of 33.3 units. Fish show the basic metameric structure of vertebrates more evidently than do the tetrapods; the trunk consists of a series of rather similar segments. Blood flows into each of these from segmental arteries branching from the dorsal aorta, and is returned in segmental veins. Each of these segmental circulations is in parallel, and all together are in parallel with the circulation of the gut and other viscera. The separate circulations of each of the gills offer another example. Not only are the eight gills of a 51 g whiting, *Gadus merlangus*, in parallel with each other but so are the 72 000 respiratory lamellae (Hughes 1966), which such a fish is likely to possess. A blood cell may pass through any one, but only one, of these in its journey round the body.

In addition to reducing the TPR, the arrangement has two other advantages. Firstly, each vascular bed is assured of a supply of fresh blood from which nothing has been withdrawn and to which nothing has been added. Secondly, the 'in parallel' arrangement facilitates circulatory regulation. Restricting flow to one or more vascular beds makes blood available for other regions. As we shall see in Chapters 10 and 11, maintaining a relatively constant arterial pressure and decreasing the resistance to flow into active tissues has been the strategy for circulatory control adopted by all fish, and indeed by other vertebrate groups.

There are circumstances when 'in series' systems are of value; they occur when one vascular bed requires blood to which some agent has been added or removed by a previous one. The gills are in series with all the other tissues of the body and fish differ from all the land vertebrates in this respect. Thus the heart has only a single ventricle, and the blood is not returned to have its pressure raised a second time, as it is in reptiles, birds and mammals. Oxygen is added to the blood prior to its passage to the tissues, but between a half and a third of the blood pressure generated by the heart is lost in overcoming the resistance of the branchial circuit. The hepatic portal circulation carries the soluble products of digestion such as amino acids and glucose forward to the liver to be reformulated into proteins and glycogen. The renal portal circulation is intercalated beween the vascular beds of the post abdominal trunk, and the capillary vessels of the kidney tubules. Through these, divalent ions such as Ca^{2+} and Mg^{2+} are excreted and necessities such as proteins and glucose are retrieved. In these situations we must surmise that the advantages of these localized transport systems outweigh the loss of pressure which resistances in series impose.

Pressures on land and in water

Measurements of blood pressure in terrestrial vertebrates are normally expressed as so many units of pressure, mm Hg or cm H_2O, above a set or reference point, which is the junction of the ventricle with the aorta. The question may be asked, what is the equivalent point in a fish immersed in a tank of water? In Figure 1 a simple manometer has been connected to the dorsal aorta of a fish. If the heart had ceased pumping, the pressure of the surrounding water, pressing on the tissues and the blood vessels in them, would force the blood up the manometer tube only as far as the surface of the water (Figure 1*A*). If, through the heart's action, a pressure of 10 mm Hg were developed (Figure 1*B*), the blood would rise up the tube by this amount: conversely (Figure 1*C*), if a vessel developing a sucking pressure of -10 mm Hg had been cannulated, the level in the manometer would be pulled below the water surface. This relationship remains valid regardless of the depth of the fish in the tank: when calibrating the blood pressure of a fish, the zero set point is the surface of the water. Flow through the blood vessels of fish in the ocean depths is not affected by the overall high pressure to which they are subjected. It is true that high pressure, like high temperature, increases the thermodynamic activity of physical and chemical systems; the hearts of deep sea fish, which cease beating at surface pressures, are restored to near-normal activity in pressure chambers at 80–120 atm (Pennec *et al.* 1988). The mechanism is not yet understood; perhaps pressure stabilizes the membranes of the cardiac muscle cells. But in blood vessels it is the relatively small differences in total energy along their length that effect flow through them, not the absolute pressure imposed by the water around the fish.

Figure 1. The measurement of blood pressure in fish; a vessel has been cannulated. In a dead fish *A*, the blood rises to the level of the water and the manometer registers zero pressure. In a live fish *B*, a positive pressure of 10 mm Hg is registered. In another vessel *C*, a sucking or aspiratory pressure of -10 mm Hg is present.

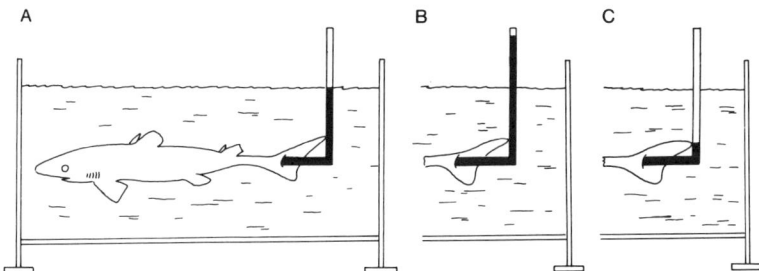

Terrestrial animals, and such air-breathing fish as emerge onto land, face the problem that blood has weight and increases the transmural pressure in the vessels of the dependent parts (Figure 2*A*). The greater transmural pressure in these vessels increases the volume of blood held in them. Brief periods of weightlessness increase the central venous pressure of man from 2.6 to 6.8 mm Hg (Norsk *et al.* 1987). Fish in water are spared this problem for they are surrounded by a liquid with a specific gravity very similar to that of blood, and the water pressure, (Figure 2*B*, *C*), acting through the tissues, exerts an equal and opposite pressure on the vessel wall. The vessel is thus not caused to dilate, as it would be in a land creature, and the delicate venous sinuses that surround, for example, the eye, do not change in volume as the fish changes from a head-up to a head-down position when feeding from the bottom. In an eel, held out of water by its jaw, the circulation fails because the blood in its central veins, lacking the support of the water, drains into the lower part of the body and does not reach up to the level of the heart.

Figure 2. The influence of gravity on blood vessel dilation on land and in water. *A*, A fluid-filled elastic tube in air. *B*, A similar tube immersed in water in a steel pipe. *C*, A fish feeding from the bottom; the numbers indicate units of pressure at increasing depths.

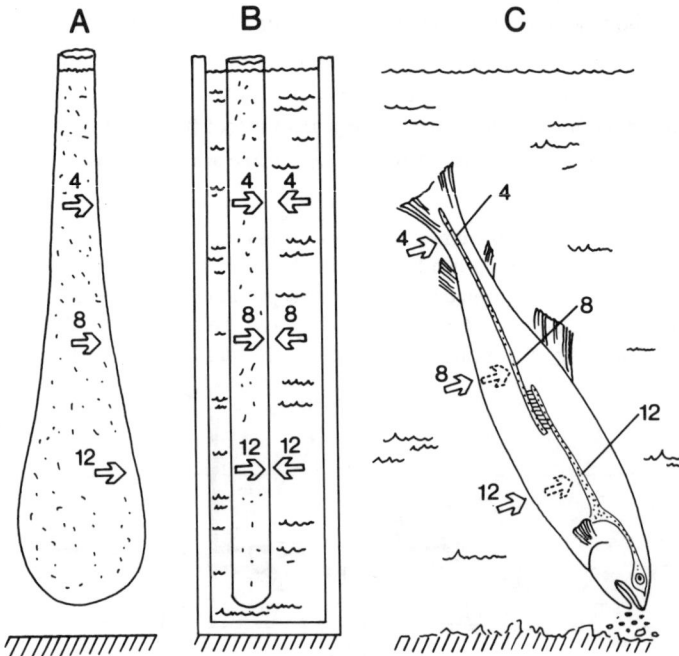

Velocity and cross sectional area

The flow of blood, measured in ml min^{-1}, issuing from the heart into the ventral aorta, is destined to pass in turn through the four segments of the circulation, the arteries, arterioles, capillaries and veins. In a 2 kg cod (*Gadus morhua*), this flow would be likely to be about 35 ml min^{-1}. If the fish is in a steady state this volume of blood must pass through each segment in a minute; it is termed the volume flow for the segment. In both the gills and the tissues, relatively few large arteries divide into more numerous small arteries and arterioles and very many capillaries; the total cross sectional area of the blood vessels of the segment increases greatly at this last division. It diminishes again, both in the vessels efferent to the gills, and in the great veins, as the many narrow vessels coalesce to form fewer, larger ones.

Velocity of flow through a segment is inversely proportional to its total cross sectional area; the 35 ml flowing each minute, through the gills or peripheral capillaries of the cod cited above, are shared amongst many thousands of channels and the blood flows very slowly indeed through any one. An analogy can be drawn with a mountain stream which includes, in its course down to the plains, two lakes, one near the summit and one half way down. The water flows slowly through the large cross section of the lakes but speeds up as it traverses the narrow gorges between. It is the cross sectional area of the whole segment that determines the velocity of flow through its vessels, not the calibre of the vessels that make it up. Hughes (1966) has calculated that the velocity of flow of the blood in the gill lamellae of the mackerel (*Scomber scombrus*) is 0.036–0.0973 cm sec^{-1}; in the ventral aorta it would be likely to be several cm sec^{-1}. This low velocity provides time for the diffusional exchanges which occur in the gills and peripheral capillaries, to approach completion.

Flow through the circulatory system of a fish is effected by a variety of propulsive mechanisms. The heart of a fish is sometimes termed the branchial heart, to distinguish it from certain auxiliary hearts and propulsers. We will term it the heart and use adjectives such as caudal and portal to designate the minor ones. The heart is certainly the outstanding organ of the fish circulatory system and provides most of the energy necessary to make the blood flow. Its structure and function will be considered in Chapter 2.

2

THE HEART

Introduction

The heart is a pump composed largely of one particular kind of tissue, cardiac muscle. The hearts of fish (Figure 3*A, B, C*) consist of three or four chambers arranged in a single series. All fish have a sinus venosus into which the returning blood flows; it may have a layer of cardiac muscle, but in many teleosts (Figure 3*C*) this is reduced to a few scattered fibres and in others, muscle is lacking altogether. The atrium is a capacious, thinly-muscled sac which generates just sufficient pressure to fill the thickly-muscled ventricle. This is most commonly sac-like, but may be pyramidal, as in the Clupeidae, or tubular as in the hake (*Merluccius merluccius*) (Santer *et al.*1983). It is the chief pressure-raising chamber of the heart and comprises 58–85% of its weight. The fourth chamber, in elasmobranchs and some primitive bony fish, is the conus arteriosus (Figure 3*B*); it is barrel-shaped and invested with contractile cardiac muscle. In most teleosts it is replaced by an elastic chamber, the bulbus arteriosus (Figure 3*C*), the wall of which contains much elastic tissue and some smooth muscle.

The efficient operation of such a heart depends on the sequential activation of its chambers. At the start some tissue must have the ability to generate a heart beat, for the fish heart, like that of other vertebrates, is myogenic. It is not activated by motor neurones like the heart of a crab. Furthermore, blood has mass and takes time to flow from one chamber to the next. Some mechanism to delay the forward march of excitation must be provided whilst flow occurs. The energy expended in accelerating the blood will increase as the square of the velocity of flow from one chamber to the next, and will ultimately be lost to the system as heat. A rapid acceleration and deceleration of blood from atrium to ventricle may be wasteful of energy.

A further problem arises from the fact that inactive muscle is easily stretched. A spherical chamber, activated only in its basal third, might waste much of its contraction in dilating the as yet inactive muscle ahead. Performance will be improved by the provision of fast-conducting tracts which travel across the wall and carry the cardiac impulse to all parts of it. In this way a near-synchronous activation can be achieved. We thus will need to consider not only the basic structure and function of cardiac muscle in fish, but also whether, and how, it is specialized to provide for these additional roles of impulse generation, delay and fast conduction.

The structure of cardiac muscle

Fish cardiac muscle, like that of other vertebrates, consists of fibres, which branch and rejoin to form a meshwork. Each fibre consists of many separate contractile cells, termed myocytes, which are mechanically joined together at their ends. Within each lies the contractile machinery

Figure 3. The hearts of fish. *A*, Atlantic hagfish, *B*, Spotted dogfish, *C*, Trout, *D*, The conus of the long-nosed gar. A. atrium, B.A. bulbus arteriosus, C.A. conus arteriosus, D.C. ductus Cuvieri, Sl.V. semilunar valves, S.V. sinus venosus, V. ventricle, V.A. ventral aorta. *A*, redrawn after Brodal and Fänge 1963. D, redrawn after Goodrich 1930.

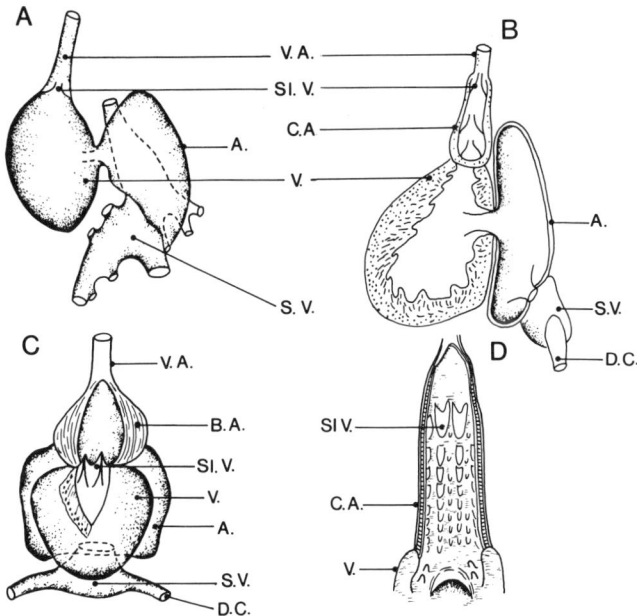

which has the same basic arrangement of sliding thick and thin filaments as occurs in skeletal muscle. Nevertheless, its fine structure differs from that of higher vertebrates in some respects and we can consider these under four headings.

The dimensions of fish myocytes

Myocytes of fish are smaller than those of higher vertebrates and contain fewer contractile elements. Canelle *et al.* (1986) give the diameter of mammalian ventricular myocytes as 10–25 μm. This can be compared with the following data from Santer's (1985) monograph; rainbow trout (*Salmo gairdneri*), 1.7–4.6 μm; common eel (*Anguilla anguilla*), 2.8–10.1 μm; albacore (*Thunnus alulunga*), 2.5–6μm; zebra fish (*Zebra danio*), 3.1–6.3 μm; spiny dogfish (*Squalus acanthias*), 2–3 μm.

Fish cardiac myocytes are often shorter than those of a mammal; an average length for those of the mammalian ventricle is 2.2 μm. Santer (1985) quotes those of spotted dogfish (*Scyliorhinus canicula*), plaice (*Pleuronectes platessa*) and perch (*Perca perca*), as 1.0–1.5, 1.4–2, and 2.2 μm, respectively. In mammalian myocytes, myofibrils are arranged in a more or less continuous mass, partly broken up by the numerous mitochondria, and penetrated by the T tubules. In many fish, myofibrils are arranged in a ring around the periphery. In the chimaera (*Chimaera monstrosa*) they occur as a single continuous tube which follows the membrane of the myocyte (Berge 1979). Leknes (1980) reports that in the atrial myocardium of a cod (*Gadus thori*) and the haddock (*Melanogrammus aeglefinus*), the myocytes contain only two or three myofibrils. Leak (1969) reports that in the Atlantic hagfish (*Myxine glutinosa*) many collagen fibres are present between the contractile elements. Some of the ventricular myocytes of inactive species like pike (*Esox esox*) contain very little contractile material (Midtun 1980). Myocytes contain many mito-chondria which have a central position; they and the nucleus occupy the longitudinal space within the peripheral ring of myofibrils (Santer and Cobb 1972, Myklebust and Kryvi 1979).

The myocytes of many fish are rich in myoglobin which is involved, through its translational movements, in the delivery of oxygen to the mitochondria. The importance of myoglobin in fish hearts has been directly tested by exploiting the difference between the hearts of the sea raven (*Hemitripterus americanus*) which possess 63.8 nmol per g ventricle of myoglobin, and the ocean pout (*Macrozoarces americanus*) which has a relatively white heart containing only 5.2 nmol g^{-1}. When isolated hearts of both species were perfused with a perfusate which had a partial pressure of oxygen (P_{O_2}) of 5 mm Hg, the myocardial power output of the sea raven heart was unchanged, whereas that of the ocean pout was reduced by 42% (Driedzic and Stewart 1982, Driedzic 1988). When the myoglobin was inactivated by an oxidizing agent oxygen extraction was similar in the two species (Bailey and Driedzic 1988).

Sarcoplasmic reticulum, caveolae and T tubules

Sarcoplasmic reticulum is less abundant in fish myocardial fibres, than it is in those of mammals. Nevertheless, its form is quite complex; immediately beneath the sarcolemma its tubules are expanded to form cisternae. Sarcoplasmic reticulum membranes are the site of a specific Ca^{2+} ATPase which in mammals may represent up to 90% of its total protein content (Carafoli 1985); it is responsible for transporting Ca^{2+} from the sarcoplasm into the reticulum. Fine tubules of it penetrate amongst, and encircle, the myofibrils. Sarcoplasmic reticulum has the key role in the homeostasis of Ca^{2+} concentration in mammalian heart cells and Ca^{2+} is released from it when the adjacent sarcolemma is depolarized. Ca^{2+} modulates a number of reactions involved in the contraction and relaxation of the myocytes and ionic concentrations of Ca^{2+} are maintained at very low levels, (μM or sub μM) inside them whilst the fluid surrounding them has levels in the mM range. In fish the store of activator Ca^{2+} is much less derived from the sarcoplasmic reticulum than it is in mammals and is to a greater extent due to inflow through calcium channels which open during the plateau period of the muscle action potential.

The sarcolemma has flask-like intuckings, < 150 nm in diameter, the caveolae. These are particularly numerous in the albacore where they are more abundant in the myofibrils of the outer compact layer of the ventricle (Breisch *et al.* 1983). Caveolae are adjacent to the subsarcolemmal cisternae and it has been postulated that they provide a primitive mechanism for excitation–contraction coupling, but, as Santer (1985) points out, there is no evidence for this. The fish myocardium, like that of the lower vertebrates in general, lacks T tubules, those deeply-penetrating tubular extensions of the sarcolemma which, in the large fibres of the mammalian ventricle, carry the muscle action potential down to the deeply-lying myofibrils. In fish, we must suppose, the Ca^{2+} which flows into the myocyte during the plateau period of the ECG, is directly available to the myofibrils lying beneath, and reaches the deeply-lying ones more slowly, by diffusion. The small diameter of fish myocardial fibres, and the tendency for the myofibrils to be arranged in a single peripheral ring, may account for the adequacy of such a simple system of coupling.

Intercalated discs, fasciae adherentes and gap junctions

If myocytes are to contract in a coordinated way, they require to be coupled together both mechanically and electrically; intercalated discs provide a mechanical coupling, and gap junctions permit an easy transfer of the cardiac muscle action potential from one myocyte to the next. In fish intercalated discs are much less convoluted than are those of mammalian cardiac muscle. Like them, however, they consist largely of fasciae adherentes. Actin filaments from adjacent cells insert onto a dense fibrous mat and the mats of the two adjacent cells are separated by modified cell membranes with a proteinaceous material between, which binds them together. Similar but smaller structures are called maculae adherentes. Scattered examples of both occur on the lateral surfaces of fibres, but fasciae adherentes are largely confined to the transversely oriented intercalated discs.

Gap junctions are regions of low electrical resistance. Freeze-fracture studies (Shibata and Yamamoto 1977) show that in the Japanese lamprey (*Entosphenus japonicus*), they consist of patches of small round or polygonal particles which bind the membranes closely together; their structure is like that of higher vertebrates but there are far fewer of them. They do not occur in the intercalated discs, mixed in amongst the fasciae adherentes (Cobb 1974), as they do in mammals, but on the lateral walls of the fibres. Shibata and Yamamoto (1977) calculate that in the trout they occupy only 0.7% of the circumference of the myocyte, compared with 5% in the fibres of the mammalian ventricle

Specific granules

Early accounts of the hearts of lampreys and hagfish (Jensen 1961, 1965, Bloom *et al.* 1961, Helle and Storesund 1975) drew attention to a layer of cells beneath the endocardium, containing electron-dense, 100–300 nm, inclusions. In lampreys they are abundant in the sinus venosus, in hagfish, in the ventricle. They have been variously termed atrial specific granules, myocardial granules and specific heart granules; similar inclusions are found in the atria of man and other mammals. In the atlantic hagfish, Helle and Storesund (1975) report a direct ultrastructural connection between them and the endoplasmic reticulum. As the endocardium, the lining layer of the heart, is in some places very thin, and its cells are sometimes fenestrated (Yamauchi 1980), the granule-containing cells are there separated from the blood flowing through the heart only by a delicate barrier, and agents liberated from them have easy access to the circulating blood. Subsequent studies showed that such cells are abundant in the sinus of elasmobranchs and occur also in the sinus and ventricle of various teleosts such as the medaka (*Oryzias latipes*) (Lemanski *et al.* 1975), cod (Leknes 1981) and plaice (Santer and Cobb 1972). In Chapter 9 we will have reason to take note of a circulating hormone, atrial natriuretic peptide ANP, which is concerned in regulating the volume of the blood and can increase the excretion of salt and water by the kidney. In man, atrial specific granules react to antisera raised to the ANP molecule and most of the stored peptide in the granules is in the 28 amino-acid form (Skepper *et al.* 1988). As we now know that fish hearts contain ANP, it is presumed that here, too, the specific granules are storage sites for the hormone, and, as in mammals, it is the stretch of the atrial myocyte that leads to ANP release (Agnoletti *et al.* 1989).

To understand the functional significance of these structural features it is necessary to emphasize that the gap junctions, sarcolemma, subsarcolemmal cisternae, and the contractile mechanism are links in a chain of excitation and are presumably matched, the one to the other. The paucity of gap junctions implies a high electrical resistance between myocytes which might be expected to slow the conduction of the action potential from one to the next. But this might, in turn, be offset by the lower electrical capacitance that must result from the small fibres and the absence of T tubules. The time constant of a resistance–capacitance network is the product of R and C, and it thus may not be greatly lengthened. Large muscle fibres have greater conduction velocities than small ones, but this may be unimportant in view of the limitations caused by the absence of T tubules. Cisternae are limited to a single layer beneath the sarcolemma and Ca^{2+} enters the fibre from outside. This in turn may have a compensating advantage in that less of the ATP generated by the mitochondria has to be used to power the Ca^{2+} pump, and more is available for the contractile mechanism. Hearts evolve as functioning wholes and we do not yet know in any detail how the different links in the chain of excitation are matched the one to the other.

Pace-makers and pace-maker tissue

Pace-maker cells can be identified physiologically because, when an intracellular recording of their membrane potential is made with a microelectrode, a pace-maker potential is registered (Figure 4*A*). Following an action potential, the resting membrane potential is briefly restored, but a steady depolarization moves this towards the trigger level at which the next action potential is fired. This then spreads through the myocardium. In mammals such cells are normally to be found in a crescent of tissue, the sinoatrial node, at the junction of the superior vena cava and the right atrium. They are histologically distinct from cardiac muscle fibres, for they are smaller, have faint or no striations and are enmeshed in a network of connective tissue. If the sinoatrial pace-maker is damaged, pace-maker activity may arise in cells in the atrioventricular node, and in yet more adverse conditions myocardial cells in the atrium and ventricle may become foci of pace-maker activity.

We are less certain of the location of the pace-maker in the fish heart, partly because it is not the same in all the many species. Arbel *et al.* (1977) note that in the African lungfish (*Protopterus ethiopicus*) the primary pace-maker occurs at the junction of the main veins with the sinus venosus. Saito (1973) has recorded pace-maker potentials from cells in the sinoatrial junction of the carp (*Cyprinus carpio*). In the Japanese loach (*Misgurnus anguillicaudatus*), Yamauchi *et al.* (1973) report that the sinoatrial node is a cushion of tissue in the sinoatrial canal and consists of modified myocytes. Its cells stain only lightly with toluidine blue, lack the

Figure 4. Cardiac muscle potentials. *A*, Pacemaker and action potentials from pace-maker cells of carp sinoatrial tissue. *B*, Hyperpolarization of plaice atrial pacemaker cells following vagal stimulation at 40 Hz. *C*, Action potentials recorded intracellularly from carp ventricular fibres. Redrawn, *A* after Saito 1973, *B* after Cobb and Santer 1973, *C* after Huang 1973.

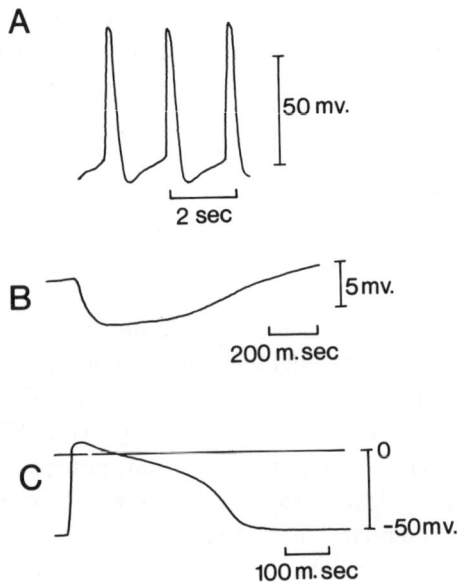

numerous specific granules which abound in the nearby atrial cells and have fewer striations. They are densely innervated by postganglionic vagal nerve fibres, the terminals of which contain the clear 30–50 nm vesicles characteristic of acetylcholine. A similar arrangement occurs in the trout (Yamauchi and Burnstock 1968). In the plaice (Santer 1972) and cod (Saetersdal *et al.* 1974) these endings have been shown also to contain acetylcholinesterase.

Saito (1973) reports that vagal stimulation hyperpolarizes the sinoatrial pace-maker cells of the carp (Figure 4*B*); this increases the resting membrane potential and a longer period must intervene before the pace-maker potential once more carries the membrane potential down to the trigger level. The heart thus beats at a slower rate. However, in the plaice, Cobb and Santer (1973) failed to find pace-maker cells at the sinoatrial junction, despite a careful search, but did find isolated cells widely scattered throughout the atrial wall. In these, too, vagal stimulation caused hyperpolarization and delayed the onset of the next action potential. A similar finding of scattered pace-maker cells is reported by Laurent (1962) in the atrium of the brown bullhead (*Ictalurus nebulosus*), and by Jensen (1965) in the atrium and ventricle of the Pacific hagfish *Eptatretus stouti*.

The atrioventricular and ventriculoconal delays

The delay of the cardiac impulse at the atrioventricular junction provides time for the ventricle to fill. In higher vertebrates, atria and ventricles are separated by fibrous rings which prevent the direct passage of electrical activity from one to the other. Only at the atrioventricular (A–V) node is a conducting pathway available. It is a narrow band of modified myocytes on the dorsal or posterior surface of the heart. In mammals most of the atrioventricular delay occurs in the first millimetre of the node, adjacent to the atrium. A–V nodal tissue itself generates pace-maker potentials but the rate at which it does so is only some two thirds that of the sinoatrial pace-maker. As the region of the heart having the highest rate of rhythmicity determines the rate of heart beat, the sinoatrial node becomes the effective pace-maker. At the A–V node there is normally an interaction of the brief propagated excitatory potential arriving from the atrium, with a resident pace-maker potential of slower rate. The vagi, it is now believed, play a vital role in inhibiting the emergence of subsidiary pace-makers and maintain a S–A node domination of the cardiac rhythm (Vassalle 1977).

We are less certain about the details of A–V conduction in fish. In the plaice, Santer and Cobb (1972) note that the atrium is separated from the ventricle by a fibrous partition and a band of small diameter myocytes forms a bridge between the two chambers. Presumably these fibres have a low conduction velocity, and this provides the delay. Laurent (1962) found that in the brown bullhead, nerve fibres from the atrium extend towards the A–V junction but the ventricle itself is poorly innervated. A–V delay is shortened by a rise in temperature and by catecholamines (Peyraud-Waitzenegger *et al.* 1980). Ripplinger and Pierron (1973) report that in the tench (*Tinca tinca*), cutting both vagi abolishes A–V coordination and in fish as in higher vertebrates, autonomic activity may be continuously necessary for the coordinated contraction of the cardiac chambers. Certainly there is a clear delay between the activation of the atrium and the ventricle and, as we shall see, in the Atlantic hagfish this delay is so long that we can observe events in the electrocardiogram not normally visible in other vertebrates (Chapter 12).

In the elasmobranchs and primitive bony fish, a ventriculoconal junction separates the muscular conus arteriosus from the ventricle. Nodal tissue is present, and the electrocardiogram shows a delay between the deflections derived from the two chambers. We know very little about this junctional tissue, but an isolated conus with its attached nodal tissue will contract rhythmically.

Fast conducting tracts

In mammals, myofibres specialized for fast conduction pass from the A–V node, down the septum, to all parts of the ventricular myocardium. These tracts, the Bundle of His and the Purkinje fibres, differ from normal myocytes. They are of larger diameter, have fewer myofibrils, mostly lack T tubules and have a high content of glycogen. In fish no such specializations are known to occur and fast-conducting tracts have not been identified histologically. However, they certainly occur: Ripplinger and Pierron (1973) studied the passage of electrical activity over the ventricle of the tench and found that it first appears on the dorsal surface, around the A–V junction, and spreads rapidly at this level around the left side of the heart to the ventral surface and down a submedian line towards the apex; from these positions it propagates more slowly through the remainder of the ventricle and finally dies out in the ventriculobulbar region close to its point of origin, some 60–100 msec later. Chiesa *et al.* (1962), in a study of ventricular activation in trout, eel and carp, found that the left side of the apex is activated before the right. Arbel *et al.* (1977), in their study of the heart of the African lungfish, concluded that whilst no specialized conducting tracts could be recognized histologically, excitation followed along set pathways in the ventricle.

Electrical properties of fish cardiac muscle
The resting membrane potential and muscle action potential
Cardiac muscle in fish, like vertebrate muscle everywhere, maintains across its cell membranes a resting membrane potential; values determined by impaling fibres with microelectrodes are: carp atrium 57.4 mV, carp ventricle 54.9 mV, yellowfin croaker (*Umbrina roncador*) ventricle, 59.6 mV (Jensen 1965). These values are below those of higher vertebrates; even the ventricular fibres of the leopard frog (*Rana pipiens*) have a resting membrane potential of 74.8 mV.

Whereas in fish skeletal muscle, an action potential is a brief affair lasting a few milliseconds, during which the resting potential is reversed, in cardiac muscle it is much more prolonged (Figure 4*C*). The initial spike-like part of the action potential lasts for 12–200 msec and is followed by a plateau period of 200–30 msec. The membrane potential during the plateau is at or close to zero and the heart muscle at this time is refractory to further stimuli. The period corresponds to the time when the muscle fibre is developing tension. As the plateau period draws to a close, the resting membrane is restored. The initial spike-like portion of the action potential of the opaleye (*Girella nigricans*) peaks at 75.9 mV and it has a reversal or overshoot of 32.7 mV (Jensen 1965); the overshoot, in yellowfin croaker, 37 mV, is even greater.

These potential changes accompany changes in the permeability to specific ions. The action potential commences with an increase in the permeability to, and an inrush of, Na^+; the membrane potential is finally restored by an outflow of K^+. During the period of the plateau Ca^{2+} channels open and allow an influx of Ca^{2+} from the intercellular spaces into the myocytes. In the hearts of fish this inflow is the primary source of the Ca^{2+} which attaches to binding sites on the troponin molecules of the filaments and activates the contractile mechanism.

Both the amplitude of, and the rate of rise of, the cardiac muscle action potential are dependent on the ratio between the external and internal concentrations of Na^+ and an increase in this ratio increases both. In the cardiac muscle of most vertebrates there is a strict limit to demonstrating this relationship by adding a sodium salt, because the external saline becomes grossly hypertonic. Elasmobranchs differ from most fish in that they are isosmotic with sea water and almost half of the osmotic pressure is provided by urea. By substituting Na^+ for urea Seyama and Irisawa

(1967) have shown, in the Japanese skate (*Dasyatis akajei*), that there is a linear relationship between the external concentration of Na^+ and the rate of rise of the action potential.

The electrocardiogram (ECG)

When the membranes of many thousands of cardiac muscle fibres are activated, which happens as a cardiac impulse sweeps across the heart, the active tissue becomes negative with respect to the rest and acts as a sink for current flow. Current flows in from the surrounding tissue into the active zone which sets up an electric field around the heart and if electrodes are suitably placed, this can be recorded. Such records are termed electrocardiograms.

We noted earlier that the sinus venosus often has little or no cardiac muscle and is simply an elastic collecting chamber preceding the atrium; the ECGs of such fish lack any sinus deflection. When however, the sinus is well-muscled, its activation contributes the V wave, the initial deflection of the ECG in Figure 5B. The activation of the atrium is indicated by the P wave; this has two components, Pa and Pt (Figure 5B), indicating the depolarization and repolarization of the atrial myocardium. Pt is seldom if ever seen but is evident in the heart of hagfish (Figure 3A) in which conduction between the atrium and ventricle is very slow (Chapter 12). QRS is a large deflection, for the ventricle is a considerable mass of muscle. The trace returns to the base line during the plateau time of the

Figure 5. Electrocardiograms of *A*, European catfish, *B*, New Zealand hagfish, *C*, spiny dogfish. *D*, conal electrogram of Port Jackson shark. Time marker = 1 sec. Redrawn, *A* after Laurent 1962, *B* after Davie *et al.* 1987, *D* after Tebecis 1967.

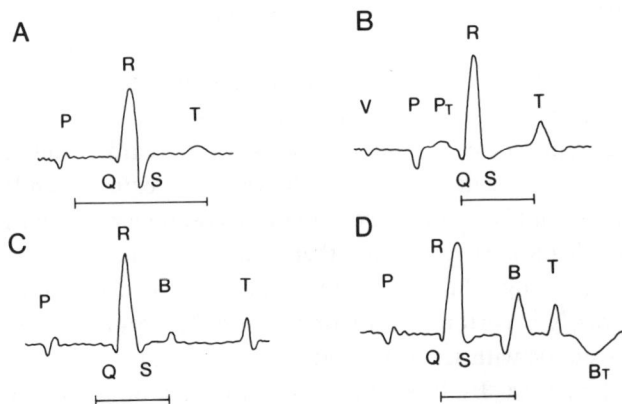

cardiac muscle action potential (Figure 5*A*); the T wave signals the repolarizaton of the ventricle. In those fish which have a conus arteriosus encircled by cardiac muscle, this may contribute to the ECG. The so-called B wave (Figure 5*C*) signals conal depolarization and occurs between R and T (Kisch 1948). As Tebecis (1967) has pointed out, there is a conal T wave, Bt, but this is only seen in a conal electrogram (Figure 5*D*) when an electrode is placed directly on the conus.

The events of the cardiac cycle

The heart is a two, three or four step, pressure-raising pump and much of its mechanism can be understood by observing the sequence of pressure changes in its chambers. These changes, often termed 'the events of the cardiac cycle' may be displayed along with the ECG and sometimes flow from the heart, and are illustrated by diagrams like that in Figure 6.

It needs to be remembered that when two chambers are in communcation they will share a common pressure profile and that the chamber which is contracting and expelling blood will have a slightly higher pressure than the one into which the blood flows. When the pressure traces intersect and part company, it suggests that the two chambers are no longer in communcation because, at that instant, the valves between them closed. We shall thus pay attention to these points of intersection of the pressure traces, for they may indicate the successive opening and closing of the sinoatrial, atrioventricular or semilunar valves.

The atrium

Figure 6 shows the events of the cardiac cycle in a trout, one of that large group of teleosts in which the sinus venosus lacks cardiac muscle. There is thus no V wave in the ECG. Following the P wave the atrium contracts and atrial pressure rises above that in the ventricle. Only during its systole does atrial pressure rise above ambient because the pericardial cavity, in which the atrium lies, itself experiences a fluctuating, subambient pressure. Inside the atrium, the delicate wall bears two radiating fans of muscle fibres, the musculi pectinati, which arise from the atrioventricular opening. When they contract they pull the atrium down onto the ventricle (Figure 7) and force the blood into it. The ventricular fibres are passively stretched as the blood flows in and ventricular pressure rises. The atrium commonly develops pressures of 2–4 mm Hg. The ventricular pressure trace suggests that most of the inflow of blood occurs early in atrial systole; shortly after this the cardiac impulse will invade the fibres of the A–V node.

The ventricle

As the cardiac impulse starts its journey along the fibres of the ventricular muscle, QRS occurs. Incipient back-flow closes the A–V valves and ventricular pressure rises rapidly as the activated muscle presses on the blood within it. After a short period of time, 0.1–0.25 sec, pressure has risen to a level at which it can open the semilunar valves which lie ahead

Figure 6. The events of the cardiac cycle in the trout. From the bottom up the traces show, pressure in P the pericardium, At the atrium, V the ventricle, and V.A. the ventral aorta. The top trace is the electrocardiogram. Redrawn from a figure supplied by Dr G. R. Bennion.

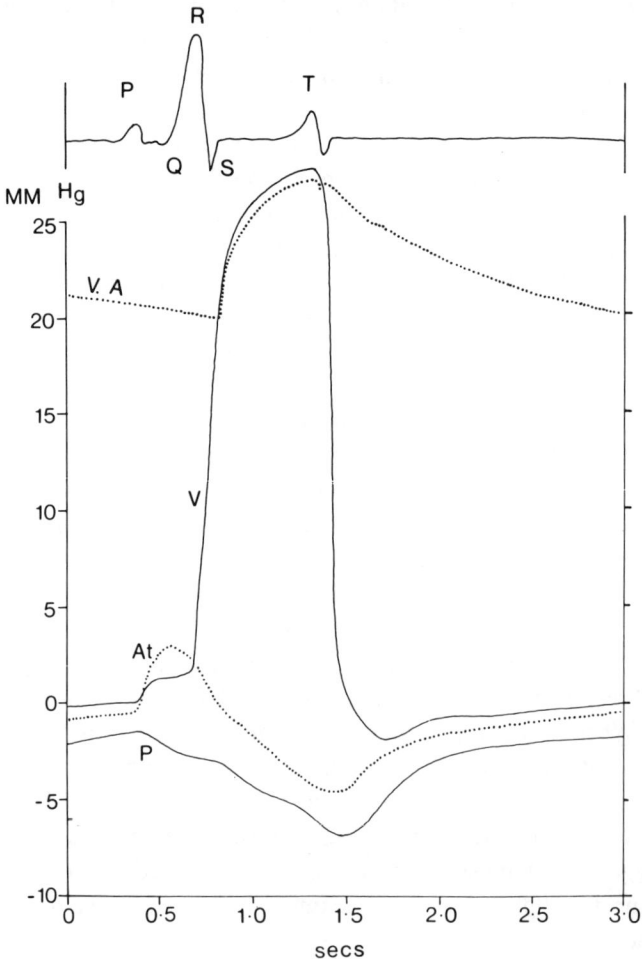

and allow blood to enter the bulbus; blood flows into and through it, to the ventral aorta. The brief period during which both the A–V valves and the semilunar valves are closed is termed the period of isometric contraction, for during it there is no change in the length of the ventricular muscle fibres. They are simply tightening onto the mass of blood within and raising its pressure towards the level of that in the bulbus and ventral aorta. Ventricular systole continues after the opening of the semilunar valves and the initial rapid outflow of blood; pressure continues to rise, although more slowly. Eventually the T wave of the ECG indicates that ventricular contraction has ceased; pressure starts to fall rapidly and incipient back-flow closes the semilunar valves. Pressure in the bulbus and ventral aorta decline steadily as blood flows through the gills to the peripheral vessels, during the remainder of ventricular diastole, until the next systole once more elevates it. The rate at which it declines, i.e. the steepness of the downward slope of the ventral aortic pressure trace, will depend on the heart rate and the peripheral resistance. A rapid heart beat and a fast run-off due to a low peripheral resistance, will result in a steeply-sloping pressure trace.

Figure 7. The heart of a teleost fish within the pericardium. A. atrium, A.-V.V. atrioventricular valve, B.A. bulbus arteriosus, Co. coracoid, C.M. compact myocardium, Pc. pericardium, Ph.M. pharyngeal muscles, Ph.Sk. pharyngeal skeleton, S.-A.V. sinoatrial valve, Sl.V. semilunar valves, S.M. spongy myocardium, S.V. sinus venosus, V. ventricle, V.A. ventral aorta.

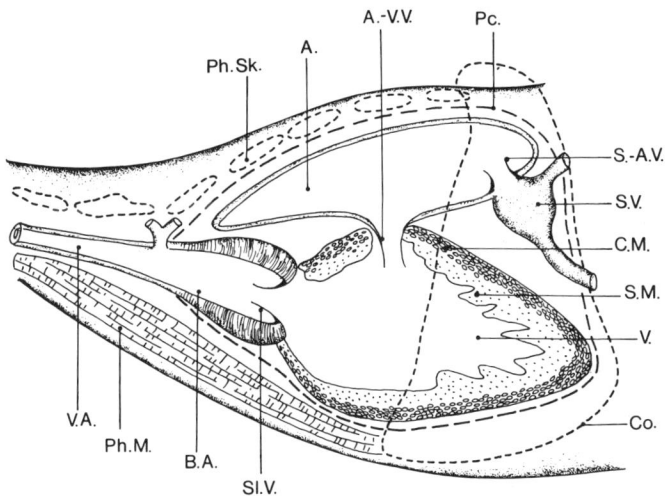

The pericardium and pericardial fluid

Attention must now be paid to the pericardium. The vertebrate heart is enclosed within a membranous bag, termed the pericardium, formed in development as a separated portion of the perivisceral coelom. It thus has an inner and an outer layer; the inner layer becomes applied to the surface of the heart, forming the epicardium. The outer layer in fish rarely forms the thin membranous bag seen in higher vertebrates, for it becomes attached to adjacent skeletal structures, such as the pectoral girdle and the basal elements of the branchial skeleton (Figure 7), and is additionally attached to and supported by the powerful pharyngeal muscles below. It thus cannot easily follow the changes in volume of the heart within it, and comes to play a significant part in cardiac filling. When blood is expelled from the ventricle during ventricular systole, pressure in the pericardium falls as its walls, made semirigid by their adherence to these structures, are tensed inwards (Figure 6). This sucking or aspiratory pressure pulls on the thin-walled atrium and in Figure 6 and 11, we can see that atrial pressure follows pericardial pressure down to its lowest level at the end of ventricular systole, when the net balance of outflow over inflow into the heart ceases to be positive. Ventricular ejection aids atrial filling, for the fall in atrial pressure steepens the gradient of pressure across the sinoatrial opening and assists the rapid inflow of blood from the central veins and sinuses.

This aspiratory pressure is termed *vis a fronte* to distinguish it from the remnant of positive aortic pressure that may persist after its decrement by the resistance of the vessels, the *vis a tergo*. Pericardial pressures in fish vary; in the albacore it is -2.6 to -7.4 mm Hg (Lai *et al.* 1987), in the common eel -4 to $+4$ mm Hg (Mott 1951), in the thornback ray (*Platyrhinoidis triseriata*) -2.1 to -2.7 mm Hg (Shabetai *et al.* 1985). The sea raven has a pericardial pressure just above ambient as have the hearts of hagfish.

The roles of *vis a fronte* and *vis a tergo* have been investigated by perfusing the trout heart in situ with the pericardium intact or opened (Farrell *et al.* 1988a). Opening the pericardium reduced cardiac output by 44%. The authors showed (Figure 8) that 65% of the maximum stroke volume was developed by *vis a fronte* filling, and that only the upper 35% depended on *vis a tergo*. The trout, it seems, operates during most of its day to day activities with an aspiratory venous return, and central venous pressures rise above ambient only during vigorous exercise. These authors calculated the myocardial power output of the heart; it is derived by multiplying cardiac output by output pressure. The pericardium facilitates interaction between the cardiac chambers; it acts as a shunt for some of the output power. By restricting ventricular contraction to a small extent and thus reducing output pressure, the pericardium shunts some of the power back from the ventricle to enhance flow into the atrium. It is this transfer which defines *vis a fronte* return, rather than the precise level of the input pressure in relation to ambient.

We see that ventricle and atrium are involved in a reciprocal relationship, for ventricular filling is achieved solely by atrial systole; there is no direct flow through from the veins as there is in a mammal. Harvey (1649)

Figure 8. The ventricular function curve of the perfused trout heart with the pericardium intact, O, and opened, □. The arrow shows the shift to the right this entails. The stippled area shows where the heart was functioning by *vis a fronte* filling, i.e. input pressures were subambient. The two pairs of blacked-in points show the power output when output pressure was raised from 37 to 52 and then to 59 mm Hg, demonstrating homeometric regulation. Redrawn from Farrell *et al.* 1988a.

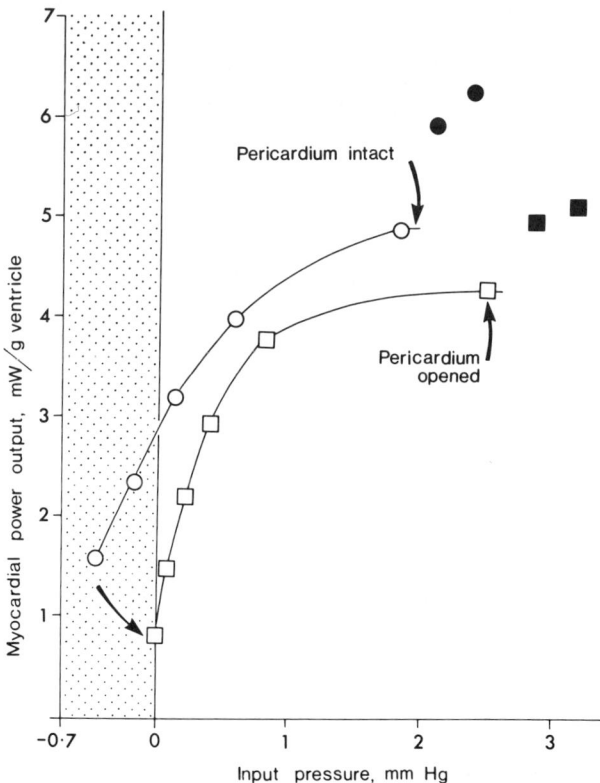

knew that in the fish heart, atrium and ventricle act reciprocally and he likened them to the action of antagonistic muscles at a joint. The contraction of each facilitates the relaxation of the other. The measure of reciprocity that exists between the atrium and ventricle in fish arises from the fact that they are both enclosed in the same semirigid pericardium. Harvey's percipience, more than 300 years ago, seems the more remarkable in view of the current interest, amongst mammalian physiologists, in the facilitating interactions of the cardiac chambers on the diastolic properties of the left ventricle. X-ray tomographic studies of the dog heart within the intact thorax show that at ambient airway pressure heart volume is virtually invariant throughout the cardiac cycle (Hoffman and Ritman 1985).

The pericardium of elasmobranchs is connected to the abdominal cavity by a narrow pericardio-peritoneal canal, the thin walls of which exert a valve-like action. When fluid is injected into the pericardial cavity, pericardial pressure rises steadily to attain a plateau at 0 mm Hg (Shabetai *et al.* 1985). Radioisotope labelling of the infused fluid shows that it passes through the canal, into the abdomen, once the plateau is reached. The canal is too delicate to allow fluid to be sucked back. In both elasmobranchs and teleosts a small quantity of pericardial fluid is normally present and the amount of it must affect the level of pericardial pressure and the end diastolic volume of the atrium. An excess of fluid in the pericardium results in cardiac tamponade and reduces cardiac output by limiting the expansion of the atrium.

The extent of subambient pressure in the pericardium varies with heart rate. A low rate, with long cardiac cycles, allows more time for the atrium to fill and intrapericardial pressure to approach closer to, or exceed, ambient. When this happens the fish may 'cough', i.e. the pharynx contracts in an expulsive movement and the muscular compression raises pericardial pressure further, expels pericardial fluid through the canal, and once more lowers intrapericardial pressure. A fast heart rate maintains a lower pericardial pressure for the semirigid walls have less time to relax back to their untensed position, before they are once more pulled in again.

The inner lining layer of the pericardium, i.e. the epicardial bounding layer of the heart itself, may be concerned in secreting the pericardial fluid. In the chimaera (Berge 1979), and in the starry dogfish (*Scyliorhinus stellaris*) (Helle *et al.* 1983), the epicardial cells appear densely granular and their surfaces bear many small cytoplasmic extensions protruding into the pericardial fluid. Lemanski *et al.* (1975), in their account of the

Table 1. *The composition of pericardial and perivisceral fluid and plasma in elasmobranchs (in mM l^{-1})*

Fluid	Na^+	K^+	Ca^{2+}	Mg^{2+}	pH	CO_2	Cl^-	SO_4^{2-}	PO_4^-	Urea	Glucose (mg %)
Plasma	255	6	5	3	7.6	8	239	0.5	1	350	90
Pericardial	353	21	1	1	6.2	< 1	366	—	1	250	1
Perivisceral	299	5	2	15	6.1	< 1	332	14	1	350	—

Values derived from the clear ray *Raja diaphenes* and from the barn door skate *R. stabuliforis*.
After Maren, 1967.

epicardial layer of the Japanese medaka, note that its cells have numerous pinocytotic vesicles and ribosome-studded cisternae which suggest it is engaged in some secretory activity. Smith (1929) and Maren (1967) noted that the pericardial fluid differs in composition from plasma and from the perivisceral fluid. It contains more Na^+, K^+ and Cl^-, less urea, and is more acidic than plasma (Table 1). It differs again from perivisceral fluid in its lower content of Mg^{2+} and So_4^{2-} and its higher content of K^+ and Cl^-.

The difference in composition between pericardial and perivisceral fluid confirms that the pericardio-peritoneal canal, when present, does not allow back-flow into the pericardial cavity. We do not know if the secretory activity of the epicardium can be regulated; in elasmobranchs the pericardial wall has a sensory innervation derived from the fifth branchial nerve. Certainly a control of the rate of secretion of the pericardial fluid would provide a mechanism for regulating pericardial pressure and hence the level of preload.

The bulbus arteriosus

The bulbus is an elastic chamber within the pericardial cavity of teleosts; a ring of three valves lies at its junction with the ventricle. Its wall has three layers (Licht and Harris 1973); the media, composed of smooth muscle and elastic tissue, is the thickest. The bulbus wall is very elastic and an increase of pressure in the carp bulbus from 7.3 to 33 mm Hg increases its volume by 700%. In this range it is 32 times more distensible than the human thoracic aorta. Priede (1976) has calculated that in the trout it can, at its full dimension accommodate some 25% of the blood ejected by the ventricle and then release it through the remainder of diastole. Measurements with flow probes show that in the lingcod (*Ophiodon elongatus*) 30–40% of the flow occurs during diastole (Farrell 1979).

This elasticity is in part due to the internal structure of the bulbus. In the trout some 10 radial septa run the length of the chamber (Figure 9*A*); each bears radially arranged bands of elastic fibres and smooth muscle. The inner ends of these insert onto thicker longitudinal bands which form the inner ends of the septa (Figure 9*B*). Priede (1976) suggests that these serve to counteract the effect of LaPlace's law, i.e. that the tension of the wall is inversely proportional to the radius of the chamber. As the bulbus expands, the radial and longitudinal elastic fibres are stretched and the pressure maintained despite the larger radius. The elastic fibres of the bulbus also seem to have some unique biochemical properties. Licht and Harris (1973) note that they dissolve in 0.1 M NaOH and in hot concentrated formic acid, which those of the mammalian aorta do not. Lansing (1959) remarks, of the wall of the bulbus of the angler fish (*Lophius piscatorius*), that it is more elastic than any other biological material he had previously examined. The only muscle in the bulbus is smooth muscle and bands in the outer wall are arranged in longitudinal spirals joined end to end by desmosomes. These spirals are innervated, by adrenergic nerve fibres, only at their outer ends and not laterally (Watson and Cobb 1979).

The bulbus, it is believed, acts as a depulsator; it converts kinetic work

Figure 9. The structure of the bulbus arteriosus. *A*, Transverse section of the bulbus of a trout. *B*, A segment cut from the bulbus to show the arrangement of elastic fibres in two septa. *C*, Ventral aortic flow in one cardiac cycle of the cod. Ad. adventitia, In. intima, L.F.E. longitudinal fibres of elastic tissue and smooth muscle, Md. media, R.F.E. radial fibres of elastic tissue and smooth muscle, Sp. septa. Redrawn, *A*, *B* after Priede 1976, *C* after Johansen 1962.

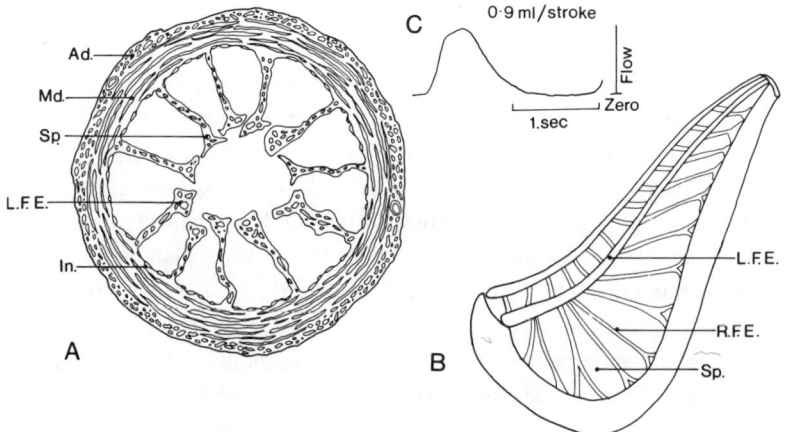

into pressure work during systole, and this is then converted back again during diastole. It spares the delicate microcirculation of the gills some of the peak of pressure caused by ventricular systole; Johansen's (1962) records of flow in the ventral aorta of the cod (Figure 9C), show that the stroke volume is more evenly spread through the cardiac cycle than is that of an elasmobranch (Figure 10). In many teleosts the atrium lies far forward in the pericardium directly above the bulbus and is grooved on its under surface to receive it. During diastole the atrium fills as the bulbus empties and the reciprocal relation, between outflow from the one and inflow into the other, must be facilitated by their close juxtaposition within the pericardium, and by the very compliant wall of the atrium. The bulbus is not, as was formerly maintained, part of the ventral aorta which has extended into the pericardium. Its different structural proteins and its unique architecture identify it, Priede (1976) suggests, as an intrinsic part of the teleost heart derived, probably, from the conus.

The conus arteriosus

The conus arteriosus (Figures 3, 10), is a muscular chamber enclosed within the pericardium, between the ventricle and the ventral aorta, in the

Figure 10. Pressures within the chambers of, and flow from, the conus arteriosus of the Port Jackson shark. Redrawn after Satchell 1970.

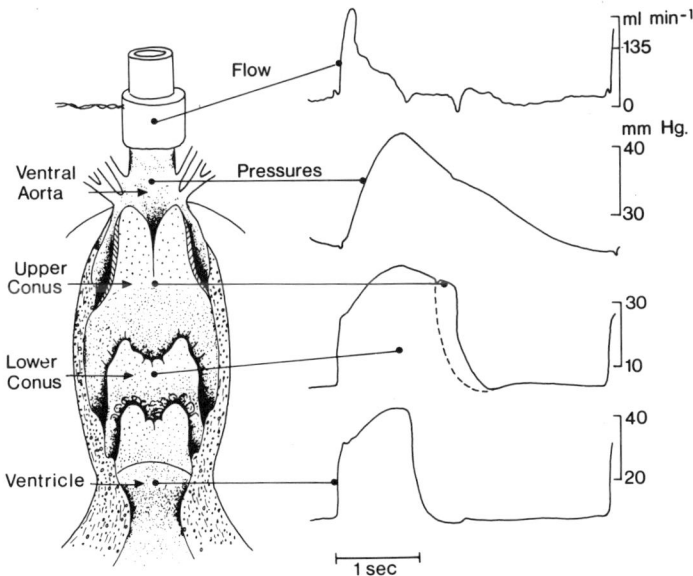

Elasmobranchi, Holocephali and the more primitive families of bony
fish. A layer of cardiac muscle encircles and overlies an elastic fibrous
wall; in most elasmobranchs two or three sets, each of three valves,
project into the lumen. The lower ones lie at the junction of the ventricle
and the conus (Figure 10), and the middle ring, when present, lies just
above these. The upper ring, at the junction of the conus with the ventral
aorta, consists of three much larger valves; only these are long enough to
span the width of the conus when it is relaxed.

During atrial and much of ventricular systole (Satchell and Jones 1967)
the lower and middle valves are open, for the conus is relaxed and the
valves cannot span the lumen. Conal pressure (Figure 10) follows ventricular pressure, for the two chambers are in communication. In the
elasmobranch heart there is not, strictly, a period of isometric contraction, for the conus dilates as the ventricle contracts. Half way through
ventricular systole the occurrence of the B wave of the ECG signals that
conal systole has commenced. This brings the basal valves sufficiently
close together so that, as the ventricle relaxes and ventricular pressure
falls, these valves close and pressure in the conus remains high. Shortly
thereafter, when conal systole has proceeded further up the chamber, the
lower valves open and the middle valves close. Finally, conal systole
comes to an end and the apical tier of large valves closes and they span the
relaxed conus for the remainder of ventricular diastole. Systole moves up
the conus quite slowly; Tebecis (1967) reported that the conduction
velocity of conal depolarization is 2–4 cm sec^{-1} in the Port Jackson shark
(*Heterodontus portusjacksoni*) and this can be compared with the
40–100 cm sec^{-1} for the movement of the QRS wave across the ventricle
in this species.

The role of the conus in cardiac function is uncertain. The flow meter
records of Satchell and Jones (1967) (Figure 10) do not support the view
that it contributes significant flow following ventricular systole. But this
may be an artefact of a study made, inevitably, on anaesthetized immobi-
lized fish. Johansen *et al.* (1966) made telemetric flow records on swimming horn sharks (*Heterodontus francisci*); these show an appreciable flow
pulse following that caused by the ventricle. The profile of dorsal aortic
blood pressure in the spiny dogfish (Satchell 1960) also suggests that there
is a small second outflow of blood from the heart following that from the
ventricle.

It is likely that the conus contributes some depulsating effect, like that
of the bulbus. Perhaps more importantly, it postpones, until later in the
cardiac cycle, the moment when back-flow from the aorta will be stopped

by valve closure, to an instant when the subambient, or sucking, pressure within the pericardium will have diminished because some blood will have returned to the atrium. The pericardium in elasmbranchs is particularly well supported by skeletal structures and the pectoral girdle is hollowed out to receive the heart. Subambient pressures as great as − 4 to − 5 mm Hg are to be recorded there, and these may interfere with a passive closure of the semilunar valves. The conus provides an active closure for the brief period in each cycle when subambient pressures are greatest. Blocking the spread of the action potential in the conal muscle leads to valvular incompetence and back-flow (Satchell and Jones 1967).

The conus is to be regarded as a primitive structure; in the lancelet (*Branchiostoma lanceolatum*), there is no single median branchial heart, but the whole of the ventral aorta, and other blood vessels, are contractile. The primitive holostean, the long-nosed gar (*Lepisosteus osseus*), has an elongate conus (Figure 3*D*), containing 48 valves arranged in 10 rows. The primitive chondrostean, the bichir (*Polypterus*), also has a very long conus. It appears to be a region of the heart that has been progressively lost in evolution, and some primitive teleosts such as the Atlantic tarpon (*Megalops atlanticus*) still retain a vestige of the conus below the bulbus.

The sinus venosus

Few fish that have a muscular sinus venosus have been investigated. Chow and Chan (1975) have recorded pressure changes in the sinus and atrium of the Asiatic eel (*Anguilla japonica*), and Davie *et al.* (1987) have related sinus pressure to the ECG, in the New Zealand hagfish, (*Eptatretus cirrhatus*). The rise in sinus pressure in the eel has an initial gentle slope (Figure 11), reflecting the slow return of blood to the sinus under the aspiratory pull of the subambient pericardial pressure. This is followed by a more abrupt rise, following the V wave of the ECG, as sinus systole occurs. The total excursion of the sinus pressure trace is only 1–2 mm Hg.

The myocardium and its blood supply
The compact layer

The wall of the ventricle in many fish is divided into an outer compact layer and an inner spongy layer (Figure 7). The compact layer comprises 27–30% of the ventricle in species such as the blue marlin (*Makaira nigricans*) but may be even more abundant (35–40%), in the swordfish (*Xiphias gladius*) and in tunas (Tota 1983). It consists of the bounding epicardium, an outer layer of longitudinal muscle fibres and an inner layer of circumferentially arranged fibres at right angles to these (Santer and

Greer Walker 1980). The spongy layer, which is present in all fish hearts, is formed of many irregular enclosures in which only the outermost muscle fibres run circumferentially. The many irregular chambers inter-communicate to form a spongy mass of spaces lined by muscle fibres; most of these spaces are directed radially so that their openings lie towards the outlet of the ventricle.

The coronary arteries; atherosclerosis and cardiac pathology
The compact layer, except in those fish in which it is very thin, has a coronary supply and profiles of blood vessels are seen in sections of it. The coronary arteries derive from two branches of the epibranchial loop vessels which surround the second and third gill slits. Tota (1983) recognizes two vascular networks arising from them. The hilar system stems from branches which penetrate the myocardium at the junction of the conus or bulbus with the ventricle. The extrahilar system arises from the main coronary vessels and spreads out superficially just beneath the epicardium. The extrahilar net may penetrate no further than this and in the plaice, the coronary circulation is limited to this superficial network.

Figure 11. Pressures within the low pressure chambers of the heart of the Asiatic eel. At. atrium, Pc.C. pericardial chamber, S.V. sinus venosus. Redrawn after Chow and Chan 1975.

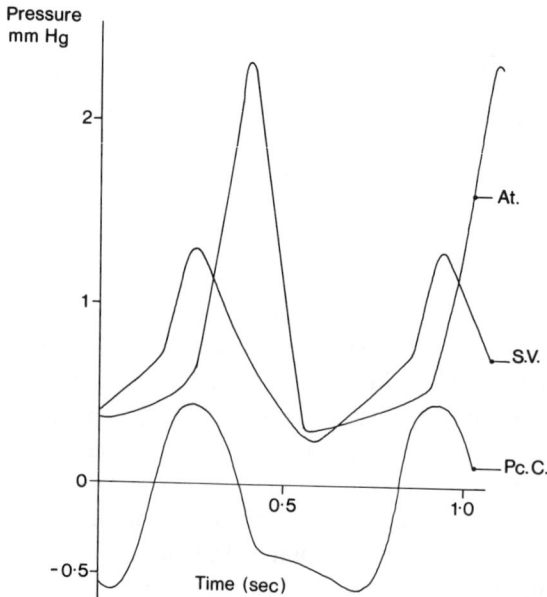

In the conger eel (*Conger conger*), the hilar vessels do penetrate the compact layer but stop short of the spongy layer. In elasmobranchs such as the spotted dogfish and torpedo (*Torpedo*) and in highly active species such as the thresher shark (*Alopias vulpinus*) arterioles from the hilar vessels give rise to capillaries in the spongy layer also, and these unite to form a system of Thebesian veins that empty into the lacunae between the trabeculae. This pattern is seen also in the ventricle of certain highly active teleosts such as the swordfish, mackerel and tuna.

Belaud and Peyraud (1971) measured the rate of flow from the coronary artery of a conger eel and found it to be only 1% of the cardiac output. This is small compred with the 4–5% in mammals, but may still be an overestimate because the outflow would be deprived of the resistance of the coronary circulation ahead. Farrell (1987) perfused the coronary artery of a trout and found that flow increased with pressure; a rise of 5 mm Hg, which can occur during exercise, would, from his study, increase coronary flow by 30%. However, myocardial power output would be likely to have increased fourfold, and the greater flow still appears insufficient. Coronary flow in fish may not be as closely matched to output as it is in mammals.

Recent studies suggest that the coronary supply is of some functional importance. The blood in it is arterialized and would be expected to have a P_{O_2} of at least 50 mm Hg. The power output of the isolated heart of the spiny dogfish perfused through the ventricle with a saline of P_{O_2} 8 mm Hg (P. S. Davie and A. R. Farrell, unpublished), was increased by 20% when the coronary arteries were perfused with an air-saturated saline. The compact myocardium of this species, i.e. the tissue perfused by the coronary arteries, constitutes 22% of the ventricle. Farrell and Steffensen (1987) found that the swimming performance of Chinook salmon (*Oncorhynchus tshawtscha*) was reduced by 30% when the coronary arteries were ligated. The adrenergic regulation of coronary flow will be examined in Chapter 9.

Salmonids show atherosclerotic coronary lesions of various severities. In the Atlantic salmon (*Salmo salar*), they may protrude slightly into the arterial lumen, or be more deeply seated and disrupt the internal elastic lamina. Severe lesions can block 18–45% of the lumen (Farrell *et al.* 1986a, 1989a). The largest lesions have the potential to increase resistance by 50% in chinook salmon and by 200% in chum salmon (*Oncorhynchus keta*). Lesions are most common in mature salmon approaching spawning, and in one sample 95% of fish had one identifiable lesion. Sixty per cent of steelhead trout (*Salmo gairdneri*) may have lesions before migration and lesions have been linked to the onset of this upstream movement. However, the lesions of post-spawned Atlantic salmon did not regress after five months in a reconditioning tank (Saunders and Farrell 1988).

The plasma lipoproteins of channel catfish (*Ictalurus punctatus*) have been studied by Smith *et al.* (1988). Three distinct classes were recognized, very low density (VLDL), low density (LDL) and high density (HDL); the last was the major lipoprotein class. Cage-reared and cultivated salmonids have high serum cholesterol levels. Sea-caged immature starved Atlantic salmon have levels of 481–619 mg dl^{-1}. Of this 61% was HDL (Farrell and Munt 1983). These values may be compared with those considered normal for man of 100–250 mg dl^{-1}. The high values have been associated with the stress of life in sea cages, and the artificial diets such fish are fed, but equally high or higher levels have been recorded from wild fish such as plaice and cod and the argentine (*Argentina silus*) (Larsson and Fänge 1977). Whether fish suffer adverse effects from these lesions is unknown.

In fish, abnormalities of the ECG suggest that they have much the same range of cardiac pathologies as man. In the laboratory, all stages of A–V block occur ranging from a simple prolongation of the P–QRS interval, to a total dissociation of these deflections. Wenckebach periodicity, in which the P–Q interval is lengthened in successive beats until one ventricular beat is dropped, has been described. There appears to be less protection against retrograde conduction, from ventricle to atrium, in fish than in mammals and such ECGs, in which an isolated QRS and T sequence is immediately followed by a P, are sometimes seen in a failing heart (Arbel *et al.* 1977). Ventricular fibrillation may occur, in which the whole surface of the ventricle undergoes a multitude of small discoordinated movements and is accompanied by a continuous series of discharges which occupy all of the electrocardiographic trace. The dramatic recovery of this condition in response to an intracardiac injection of adrenaline, and the gradual regrouping of the ECG discharges into a coordinated beat, resemble the sequence of events to be seen in man.

The spongy layer

The spongy layer, in most fish, lacks a coronary supply and coronary capillaries extend no further than the compact layer (Ostádal and Schiebler 1971). Hence, the myocytes of the spongy layer must obtain their oxygen and nutrients from the blood flowing between the trabeculae. At rest, the P_{O_2} of this may be quite high, e.g. trout, 32 mm Hg (Cameron and Davis 1970), starry flounder (*Platichthys stellatus*), 17.4 mm Hg (Wood *et al.* 1979), New Zealand hagfish, 17.2 mm Hg (Wells *et al.* 1986). In the sea raven, which lacks a coronary supply and has only a spongy layer, myocardial V_{O_2} (oxygen consumption) in a 1 kg fish with an 0.8 g heart, is 0.25 μl O_2 sec^{-1}. The venous oxygen content is 3 vol % and myocardial V_{O_2} uses less than 4% of this (Farrell *et al.* 1985). But during and after exercise, when the heart presumably needs to increase its output, and is called upon to do more work, Pv_{O_2} may fall to low levels; in the New Zealand hagfish it is, after exercise, 3.5 mm Hg.

In the trout there is evidence that the myocardial power output of the paced, isolated, perfused *in situ* heart, does depend to some extent on the P_{O_2} of the perfusate traversing the ventricle (Farrell *et al.* 1989b). In this preparation the fish was ventilating normally and the compact myocardium was thus perfused with oxygenated blood through the intact coronary system. When the P_{O_2} of the perfusate flowing through the ventricle was reduced from 67 mm Hg to 46 mm Hg the power output of 2 mW per g ventricle muscle fell by only an insignificant amount, but at a P_{O_2} of 26 mm Hg this was reduced to 1.5 mW per g ventricle and few preparations were able to perform at all, when the perfusate had its P_{O_2} reduced to 10 mm Hg, despite the continuing coronary supply to the compact layer.

Myocardial cells in the spongy layer metabolize both glucose and lactate aerobically; they can also metabolize fatty acids aerobically and in the sea raven and ocean pout a low molecular weight (12 800) fatty-acid binding protein is present (Stewart and Driedzic 1988). A study of the levels of enzymes concerned in aerobic carbohydrate metabolism relative to those concerned in aerobic fatty-acid metabolism in the hearts of higher vertebrates shows that the demand for greater power output has been met by an expansion of their ability to metabolize fatty acids aerobically (Driedzic *et al.* 1987).

The spongy myocardium may have to depend increasingly on glycolysis during hypoxia and severe exercise, and earlier workers anticipated that the compact and spongy layers would, on investigation, prove to have very different activities of the key enzymes associated with aerobic metabolism, such as cytochrome oxidase and with glycolysis, such as lactate dehydrogenase.

This expectation was reinforced by the distribution, amongst different fish, of the compact layer and its associated coronary vessels. All elasmobranchs have a compact layer but in a survey of 93 teleost fish, Santer and Greer Walker (1980) found that only some 20% did so. These were almost entirely fast-swimming, active species such as salmon, herring (*Clupea harengus*, mackerel, and various species of tuna. Inactive and sedentary fish such as cod, turbot (*Rhombus maximus*), plaice, flounder (*Platichthys flesus*), perch and goldfish (*Carassius auratus*) have only the spongy myocardium. In Atlantic salmon, Poupa *et al.* (1974) found that the activity of cytochrome oxidase in the compact layer was the same throughout the year, which might be expected as the arterial blood is likely to be at all times saturated. In the spongy layer activity was lower in summer, suggesting that when fish are more active, lower Pv_{O_2} levels impose a more glycolytic metabolism.

However, it must be said that subsequent research has not unequivocally supported this simple interpretation. In trout the activity of lactate dehydrogenase is the same in the two layers. The cytochrome oxidase activity of the total heart muscle of cod, which lacks a compact layer, is virtually the same as that of mackerel, which has one. Again, mackerel heart muscle has three times the lactate dehydrogenase activity of that of plaice, which lacks a compact layer (Gesser and Poupa 1974). These authors, in their 1973 paper, point out that the myotomal muscle of fish, particularly the great mass of white muscle, liberates lactate into the blood and the 'in series' arrangement of the fish circulation ensures that this perfuses the spongy layer of the heart the more so because the fish liver has a rather poor ability to metabolize lactate. They suggested that the biochemical adaptations of the myocardium may also reflect the proportions of red and white muscle in the trunk myotomes. Lanctin *et al.* (1980), in a study of the perfused heart of the brook trout (*Salvelinus fontinalis*), reported that lactate oxidation rates were higher than glucose oxidation rates when each substrate was provided individually, and that glucose oxidation was more strongly inhibited by lactate, than lactate oxidation was inhibited by the presence of glucose. These findings suggest that the spongy myocardium preferentially metabolizes lactate.

The output of the heart

The cardiac output of a fish is a statistic, expressed in ml min^{-1} kg^{-1} which tells us how much blood is pumped per minute. Physiologists try to measure cardiac outputs on animals at rest when their metabolic rates are at basal levels. This is not practicable for fish like tuna and mackerel which depend on forward swimming for ventilation, and the values quoted for them are likely to be high compared with species in which such measurements can be made on resting fish.

Many cardiac outputs have, in the past, been measured by the Fick

Table 2. *Cardiac output in fish*

Species	Output (ml min^{-1} kg^{-1})	Temp (°C)	Authors
Mackerel *Scomber scombrus*	54–109	—	Hughes 1966
Kawakawa *Euthynnus affinis*	68	25	[a]
Trout *Salmo gairdneri*	36.7 ± 3.9	15	Wood & Shelton 1980a
Cod *Gadus morhua*	17.3 ± 1.0	10	Axelsson & Nilsson 1986
Sea raven *Hemitripterus americanus*	12.0	10	Farrell *et al.* 1983
Lingcod *Ophiodon elongatus*	10.9 ± 0.2	10	Farrell 1981
Eel *Anguilla australis*	10.4 ± 0.63	16–20	Hipkins 1985
Spotted dogfish *Scyliorhinus canicula*	9.9 ± 0.7	14	Short *et al.* 1977
Atlantic hagfish *Myxine glutinosa*	8.9	10	Axelsson *et al.* 1990

[a] D. R. Jones, unpublished data; quoted in correspondence from A. P. Farrell, 1989.

method, which relates the difference in oxygen content of the arterialized and venous blood to the overall consumption of oxygen per minute. Such methods work well for air-breathing vertebrates. Recently it has been shown that some 20–40% of the oxygen used up by the gills in their metabolism, and thus entering into the total oxygen consumption, is taken directly by the epithelial cells from the water as it flows past them (Johansen and Pettersson 1981). Moreover, this proportion differs from one species to another. The error will tend to make Fick values erroneously too high and all the values in Table 2, except for the first, have been directly determined with flow probes. It is apparent that most fish have a low cardiac output compared with the 76 ml min^{-1} kg^{-1} of man. Only tuna and mackerel appear to approach human values and these two are perhaps overestimated for the reason outlined above.

Starling's law of the heart

When the return of blood to the heart of a fish increases, central venous pressure rises, the atrium and ventricle are dilated and their muscle fibres are lengthened. This results in an increase in the cardiac output, for the myocardium then performs more work; contraction is more vigorous and more prolonged and the ventricular pressure rises more rapidly and achieves a greater height. This response, which is common to the hearts of all vertebrates, was described by Starling (1918). Cardiac muscle, like skeletal muscle, contracts more vigorously if its myofibrils are first stretched; beyond a certain limit the increase ceases and the force of contraction becomes less. The initial increase occurs primarily because the degree of activation of the myofibrils by Ca^{2+} increases as muscle length increases. In addition, the amount of Ca^{2+} supplied to the myofibrils is greater in stretched muscle. The change in Ca^{2+} sensitivity is due to a change in the affinity of troponin. Cardiac muscle troponin has an inherently higher affinity for Ca^{2+} than skeletal muscle troponin, and when it is exchanged into skeletal muscle, imparts to it the greater sensitivity to stretch of cardiac muscle (Allen and Kentish 1985).

The hearts of fish are particularly responsive to the Starling mechanism. This has been shown in studies in which the sinus venosus and ventral aorta are cannulated and the response to the heart to different filling pressures (preload), and output pressures (afterload) are investigated. The preparation has the advantage that the pericardium and other investments are left intact, and such preparations produce cardiac outputs and perform myocardial work within the range of intact fish (Farrell *et al.* 1982, 1985, 1986b). A rise of 1 mm Hg in the filling pressure of the heart of the sea raven, i.e. from 0 to 1 mm Hg, increased cardiac output almost fourfold. In terms of the power output, this is an increase from 0.4 to 1.4 mW per g ventricle. The line relating cardiac output to preload pressure slopes steeply upwards over the physio-logical range (Figure 12*A*).

Most of this increase is in stroke volume, little or none in heart rate. Fish thus differ from higher vertebrates, which increase rate rather than stroke volume. This may be because most amphibian and reptil-ian hearts have two atria but only a single ventricle. In this the arterial and venous streams depend on lamina flow to maintain them separate. An increase in stroke volume might cause undesirable intracardiac mixing; an increase in rate does not (A.P. Farrell, personal communi-cation).

Homeometric regulation

A second type of intrinsic regulation of the heart also depends upon the relationship between initial length of the muscle and the force with which it will contract. As was noted above (Figures 6, 7), the semilunar valves open when ventricular pressure exceeds ventral aortic pressure and close when it falls below. Hence increase in ventral aortic pressure, i.e. after-load, results in a greater volume of blood in the ventricle and atrium at the start of diastole, an increased stretch of their walls and an increased force of contraction. Because of this, a rise in the resistance of the peripheral circulation, and the rise in pressure this causes, brings about a corres-ponding increase in power output of the heart. Farrell *et al.* (1985, 1986b) report that the lines relating afterload to power output and cardiac output are almost flat (Figure 12*B*). The phenomenon is termed homeometric regulation.

Whereas the Starling mechanism enables the heart to increase its output in response to a volume load, homeometric regulation mediates the response to a pressure load. Because, in a fish, the atrium is the site of the lowest pressures that cyclically occur in its circulatory system and is

Figure 12. Cardiac output in the isolated, perfused *in situ* heart, of the sea raven, subjected to different preloads □, and different after-loads ○. Preload scale on bottom abscissa, afterload scale on top abscissa. Redrawn after Farrell *et al.* 1985.

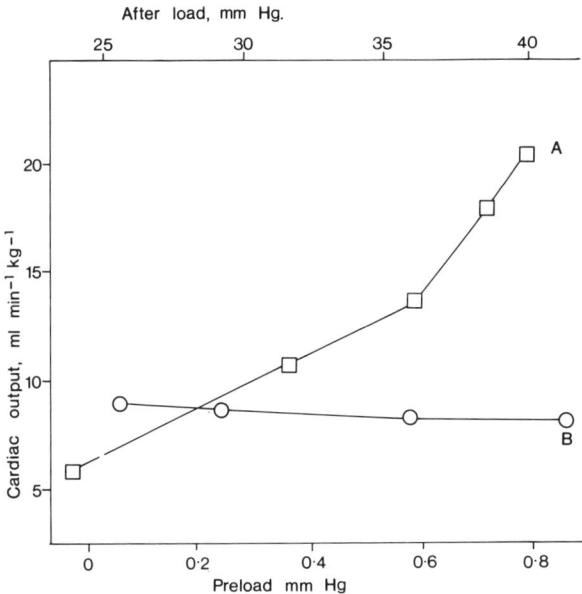

also the sole means of filling its ventricle, it comes to have this primary role in mediating the response to changes in the volume of blood returning to the heart; significantly, the cardioregulatory nerve fibres do not usually extend further than the atrium.

A third type of intrinsic response may be significant in the hearts of hagfish, which lack all cardioregulatory nerves. Chapman *et al.* (1963) and Jensen (1961, 1965) report that increased filling pressure, and hence increased stretch of the sinus venosus and atrium, elevates the heart rate. Jensen suggests that stretch depolarizes the pace-maker cells; the isolated heart of the Pacific hagfish (*Eptatretus stouti*) can increase its rate by 50–150% in response to increased filling pressure and the response occurs immediately. The isolated heart of a trout, deprived of its pericardium, can also increase its rate when perfusion pressure is increased, but this may reflect the abnormal conditions which allow the atrium to be excessively distended. No such increase in rate has been detected in the hearts of sea raven, ocean pout and trout when their hearts are perfused *in situ* (Farrell *et al.* 1985, 1986b).

Our review of the structure and function of the heart has examined its intrinsic ability to regulate the variables set out above, but it has not so far considered the effect of the autonomic nervous system, and of hormones and other agents carried in the blood, on the heart. Catecholamines, we shall see, have the capacity to alter the intensity with which the heart responds to both preload and afterload, with consequent changes in power output, stroke work and cardiac output. Cardiac function needs to be related to the circulatory system into which the blood flows; we will, in the next Chapter, examine the structure and properties of the peripheral vessels before we return to this topic in Chapters 9, 10 and 11.

3

THE PERIPHERAL CIRCULATION

Introduction

Simple dissection shows us that the circulatory system of a fish does not consist of a branching array of similar tubes, for four distinct types of vessel can be recognized: arteries, arterioles, capillaries and veins. Each performs a specific function and has a structure appropriate to this. Arteries provide for the distribution of blood to all parts of the body, with a minimum loss of pressure. Arterioles are short, narrow, muscular vessels which lower the pressure to a level at which it will not damage the delicate capillaries that lie ahead. The capillaries provide an extensive system where diffusion and exchange of nutrients and the respiratory gases can occur. The veins enable the blood to be returned to the heart and also provide a reservoir in which extra blood can be stored until needed.

Four tissue elements may be found in these vessels. All vessels are lined with a layer of thin flat endothelial cells, outside which three sorts of fibrous tissue elements may be present, each characterized by a particular protein. Elastic fibres consist of the protein elastin; when such fibres are stretched and released they tend to relax back to their original length. Collagen fibres consist of the protein, collagen; they have great tensile strength and tend to stretch very much less than elastic fibres. Smooth muscle fibres contain the proteins myosin and actin and can contract. They may be innervated so that contraction is under the control of the autonomic nervous system.

Arteries

Arteries are vessels of wide diameter and the total cross sectional area of the arterial system does not increase very much along its length so there is little loss of either pressure or velocity as the blood moves towards the periphery. The arterial wall typically consists of three layers, the intima, media and adventitia. The intima is lined with a single layer of flattened endothelial cells. Beneath this (Figure 13*A*) is a thin layer of elastic tissue termed the internal elastic lamina in which the elastic fibres are condensed to form a continuous fenestrated sheet. In fixed preparations this has a wavy outline but in life it is stretched by the pressure of the blood within. The medial layer takes up most of the wall of an artery; it consists of concentric laminae of elastic fibres interspersed with smooth muscle and a small amount of collagen. Sometimes the outer part of the media has the elastic tissue so condensed as to form an outer elastic lamina. The adventitia of the arterial wall is less well developed and consists of a thin layer of collagenous tissue.

This structure is best seen in the elastic arteries afferent to the gills. In the ventral aorta of the carp (Dornesco and Santa 1963), 20 or more superimposed elastic laminae can be discerned. This type of structure is continued into the afferent branchial and filamentar arteries. It contrasts greatly with that of the dorsal aorta. This is attached along its length to the ventral surface of the spinal column by transverse bands of collagenous fibres, and its dorsal wall, adjacent to the spine, is thin. Its lateral

Figure 13. The structure of fish blood vessels. *A*, Part of the wall of an intestinal artery of *Petromyzon marinus*. *B*, Arteriole, *C*, wall of intestinal vein, *D*, capillary, of *Squalus acanthias*. Ad. adventitia, C.F. collagen fibres, D.C.F. dense collagen fibres, E.F. elastic fibres, End.C. endothelial cell, I.E.L. internal elastic lamina, In. intima, Md. media, S.M.C. smooth muscle cell. Redrawn, *A* after Neuville 1901, *B*, *C*, *D* after Rhodin and Silversmith 1972.

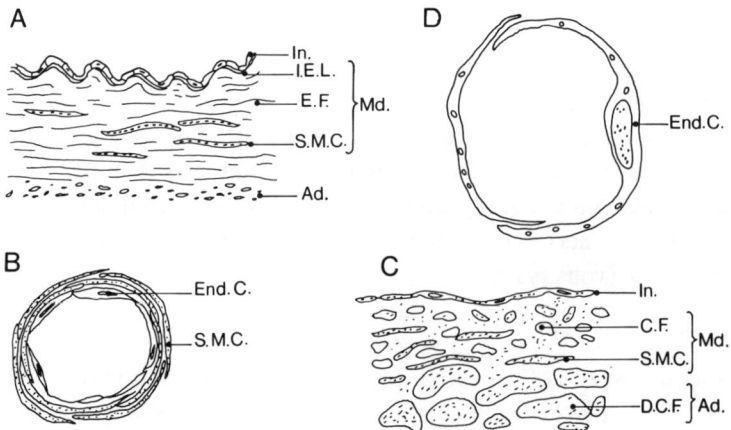

and ventral walls have a thin internal elastic lamina beneath which is some smooth muscle but the main part of the wall is densely collagenous rather than elastic.

The difference in structure between the ventral and dorsal aorta may be related to the large capacity of that portion of the arterial system in fish, which lies beyond the gills, for the dorsal aorta is the longest artery of the body. It is presumably advantageous to achieve as even a flow of blood through the gills as possible, and a sudden surge of blood through them early in each cardiac cycle, into a capacious and dilatable arterial reservoir beyond, would hinder this. Although the capacity of the dorsal aorta, due to its length, is much greater than that of the arterial vessels prior to the gills (Jones *et al.* 1974), its inelastic collagen-rich wall must be of advantage in minimizing this surge.

In some teleost fish this close association of the dorsal aorta with the spine is carried further. The spine in all fish is strengthened and converted into a supple rod by longitudinal ligaments, and in some genera the ventral longitudinal ligament indents the dorsal wall of the aorta and projects into it as a median partition. It has been described in the trout (Priede 1975), carp (Dornesco and Santa 1963), herring (De Kock and Symmons 1959), and shad (*Alosa alosa*) (Burne 1909). In the trout it completely divides the flattened aorta and Burne suggested that the sinuous movements of swimming cause it to propel the blood along the vessel.

Priede (1975) investigated this and found that in the trout the ligament is rich in elastic fibres and is under considerable tension. Experiments with mechanically simulated swimming movements in an artificial model demonstrated that such a ligament could act as a pump and that its output increased with tail beat frequency up to a maximum, after which it failed. It is not present in all teleosts and is totally absent in the pike and perch. But when it is present, Priede suggests, it may have value as an auxiliary pump which, when the fish swims, increases aortic flow.

The arteries of the intestine (Neuville 1901) are derived from branches that run from the dorsal aorta; they have a more normal structure (Figure 13*C*), with an evident internal elastic lamina and a layer of smooth muscle surrounding it. This layer, in the lamprey (*Petromyzon marinus*), has some longitudinal fibres mixed in with the circular ones.

Arterioles

Arterioles are small arteries (Figure 13*B*), with an internal diameter of 300 μm or less; they are very short, seldom more than 500 μm long, and thus tend to be overlooked in dissections. A long and distinct arteriole occurs where each side branch of the afferent filamentar artery enters a secondary lamella (Figure 19, below). Arterioles have rather little elastic tissue in them, but have a thick layer of smooth muscle. This may be innervated by the autonomic nervous system, which can change the tension developed by it. The narrow diameter of arterioles causes them to have a high resistance and in mammals, half of the total resistance of the blood system lies in the arterioles.

Capillaries

Capillaries (Figure 13D) consists of a single layer of endothelial cells rolled up to form a tube 4–10 μm in diameter and they are the smallest of all the blood vessels; our knowledge of them in fish is rather limited. In Chapter 7 we will examine structures termed retia mirabilia, and the retia of the swimbladder and eye consist of parallel arrays of arterial and venous vessels of capillary dimensions. Because of the intrinsic interest attaching to these structures, we have come to know more about retial capillaries than about those to be found in any other fish tissue. We should note, however, that they are atypical in their exceptional length, for those in the swimbladder of the common eel (Figure 28, below), may be 4 mm long. Capillaries in other tissues are 0.5–1.0 mm long.

The endothelial cells are flattened; those of the arterial capillaries of eel retia are relatively tall, 2–4 μm high, whereas those of the venous capillaries are less so, 0.2–1.0 μm high. Moreover, the venous capillaries have fenestrations closed by a diaphragm. Covered fenestrations occur also in the capillaries of the choroid plexus and median eminence of the brains of elasmobranchs, sites where the blood–brain barrier is known to be lacking in mammals. Open fenestrations tend to occur in glandular organs such as the pineal gland of the pike (Owman and Rudeberg 1970), and the gas gland of the perch (Jasinski and Kilarski 1971).

The ends of endothelial cells have clefts between them and through these are tortuous pathways which span the capillary wall; these have been revealed by ultrathin (18 nm) serial sectioning (Bundgaard 1987). Studies on arterial retial capillaries shrunk by perfusing them with hyper-osmotic solutions (Rasio *et al.* 1977, 1981), show that in places the membranes are tightly joined across the clefts by desmosomes. Rhodin and Silversmith's (1972) paper on the capillaries of the spiny dogfish shows that their endothelial cells are 1–2 μm thick and contain many pinocytotic vesicles and caveolae. Such profiles are a prominent feature in the retial arterial capillaries, and electron microscope studies show that they intrude deeply into the cell from both the luminal and basal surfaces (Bendayan *et al.* 1975). Our understanding of their role has been advanced by the protein A gold method of labelling (Bendayan 1980). This depends on coupling gold to albumin, and localizing this with anti-albumin antisera. The gold marker was seen to be concentrated in these profiles. Bundgaard (1987) suggests that they are sites of ingress of Ca^{2+} into the cytoplasm. They are not, as has from time to time been suggested, the origin of transendothelial channels (Michel 1988). They are, for example, abundant in the cells of the blood brain barrier of *Myxine*, which, as in higher vertebrates, is impermeable to most large molecules. Smooth-surfaced cisternae are present in close relation to these profiles.

Endothelial cells contain filaments of actin and myosin and can contract. In man this motility is known to be regulated by cyclic AMP and Ca^{2+}/phosphatidylserine-dependent protein kinase C (Smirnov *et al.* 1989). Forskolin, an activator of adenylate cyclase, causes rapid and reversible changes in the shape of endothelial cells, associated with the disappearance of their actin bundles. Endothelial cells secrete certain mediators. Endothelium-derived relaxing factor (EDRF) is an agent which relaxes vascular smooth muscle; EDRF, prostglandins of the E series and prostacyclin, all of which are derived from endothelial cells, are believed to be important in reactive hyperaemia, discussed below. Endothelial cells have stress-sensitive ion channels which are activated, it is believed, by mechanical forces generated by blood flow. This response to shear-stress involves an inward, K^+ selective, ionic current which hyperpolarizes the endothelial cells (Olesen *et al.* 1988). Changes in capillary permeability can occur very rapidly and, by increasing the concentration of intracellular Ca^{2+}, activate the actin–myosin filaments that occur in endothelial cells. Their contraction may increase the size of the intercellular channels (Crone 1986, 1987). These and other recent studies give a graphic picture of the endothelium as a responsive and dynamic tissue, although few of these findings have yet been extended to fish.

Underlying the endothelial layer is a basal lamina which at high magnification is seen to consist of three layers. The middle layer has many 2–5 nm fibres in it and larger 10 nm fibres occur in the outer layer (Bendayan *et al.* 1975). In the capillaries of the intestine, kidney and pancreas of elasmobranch fish, the basal lamina is anchored into the surrounding tissue by strands which extend radially outwards from it, and these may help the vessel to withstand the subambient pressures exerted by cardiac suction (Casely-Smith and Mart 1970).

Table 3. *Dimensions of the branchial vessels[a] of a 4 lb lingcod, after Farrell 1980b, and their calculated shear rates, after Graham and Fletcher 1983. Dimensions in cm, cm² and sec; shear rate calculated as 4 V/r where V = flow velocity and r = vessel radius*

Vessels	Radius	No.	Length	Flow velocity	Total cross-sectional area	Transit time	Shear rate[b]
Ventral aorta	0.2	1	3.0	5.8	0.126	0.05	116
Afferent branchial A.	0.125	8	5.0	1.86	0.393	2.69	60
Afferent filamentar A.	0.01	3760	1.44	0.62	1.18	2.33	247
Afferent lamellar arteriole	0.001	1.94×10^6	0.045	0.13	6.1	0.34	529
Lamella	0.0005	1.94×10^6	0.8	0.32	30.7	2.53	2530
Efferent filamentar A.	0.0095	3760	1.5	0.68	1.07	2.19	288
Efferent branchial A.	0.1	8	5	2.9	0.25	0.63	116

[a]The bronchial circulation is discussed in greater detail in Chapter 6.
[b]The data in this column are discussed in Chapter 4.

Capillary beds, cross sectional area and velocity of flow

The role of each component of a vascular bed, such as that of the gills or the skeletal muscles, cannot be interpreted solely in terms of the structure and dimensions of the vessels; we noted in Chapter 1 that the velocity of flow, and hence the transit time of the blood in the vessel, is inversely proportional to the total cross sectional area of all the vessels in that segment. Farrell (1980b) has measured the vessel dimensions of, and calculated the velocity of flow in the branchial vessels of a 4 kg lingcod, and some of his data are presented in Table 3.

The spaces between the pillar cells may be regarded as equivalent to the lumina of capillaries; the outermost channel of the lamella has a capillary structure, for it is lined not with pillar cells but with endothelial cells. It is clear that the ability of the lamellae to act as a diffusion and exchange system is enhanced by the prolonged transit time of the blood in them; the total cross sectional area of the lamellae, were all of them perfused, would be 244 times that of the ventral aorta. The large number of parallel channels, almost two million in the 4 kg lingcod noted above, also ensures a very large surface area across which diffusion can occur, and a slow rate of flow. The calculated transit time for the blood to pass between the pillar cells of a 0.8 mm long lamella is 2.53 sec. Hughes (1966) calculated the lamellar transit time in the mackerel as 1.0–1.9 sec.

Interchanges across the capillary wall; the Starling principle

It is widely accepted that the movement of water and dissolved substances across the capillary wall depends on a balance between three variables (Michel 1988). The hydraulic permeability, L_p is responsible for the flow of water across a unit area of capillary wall per unit difference of hydrostatic pressure. At the arteriolar end of a capillary the hydrostatic pressure is likely to be high enough to force fluid out of the capillary into the tissue spaces. Some of the substances in solution in the plasma, such as the plasma proteins, may be unable to pass through the capillary wall and will exert an osmotic pressure, termed the colloid osmotic pressure, COP. It acts to oppose L_p and causes water to pass back into the capillary. Just how diffusible a macromolecule is, will depend on its size and charge and on features of the particular capillary wall. The term diffusional permeability, P_d is a coefficient applied to a particular substance; it is defined as the mass transport of a substance per unit area, per unit of concentration difference under conditions when the flow through the wall is zero.

The extent to which proteins or other macromolecules remain in the capillary, there to exert a COP, can be defined by a third term, the reflection coefficient, σ. It is that proportion of a dissolved substance which is retained within the capillary during ultrafiltration, i.e. is reflected back into the capillary. If the membrane were completely impermeable to a macromolecule, σ would equal 1; if the solute could pass through the membrane without restriction, σ would equal 0 and the substance would exert no COP.

In man the hydrostatic pressure at the arteriolar end of the capillary is likely to be some 40 mm Hg, and thus greater than the COP of 25 mm Hg; hence fluid tends to be forced through the capillary wall into the interstitial spaces. At the venule end of the capillary the resistance of the narrow vessel will have lowered the pressure within it to some 15 mm Hg. This is less than the COP and fluid tends to move back in. There is thus a circulation of the fluid part of the blood through the tissue spaces and some 60% of the cardiac output may follow such pathways. It has to be emphasized that this hydrodynamic or osmotic flow is independent of the movement of substances by diffusion. This hypothesis of tissue fluid formation, we owe to Starling (1866–1927); it has been elaborated and given more precise expression (Michel 1983, 1988), but is still fundamental to our understanding of fluid exchange across the capillary wall. It may be asked, to what extent can this understanding by transferred to the capillary circulation of fish.

In Figure 14*A*, *B* some likely pressures in the muscle capillaries of a teleost fish are shown. Hargens *et al.* (1974) have determined the COP of cod blood as 8 mm Hg. A capillary pressure at the arteriolar end of 13 mm Hg may be surmised from the dorsal aortic pressure (Wahlqvist and Nilsson 1977, 1980) on the assumption that 50% is lost across the arterioles. Caudal vein pressures have been determined in the starry flounder by Wood *et al.* (1979) and in the winter flounder (*Pseudopleuronectes americanus*) by Cech *et al.* (1976, 1977). Hydrostatic pressure and COP in the fish capillary are seen to be in much the same ratio as they are in mammals, and we may presume that a Starling balance occurs. Gingerich *et al.* (1987) note that blood volumes of bony fish, determined with labels that adhere to plasma proteins, are 25–50% greater than those based on labelled red cells, a disparity which would be explained if a proportion of the plasma protein is normally outside the blood vessels; it implies that the σ values for the major macromolecules are less than 1 in fish, as they are in higher vertebrates.

The two pathways through the capillary wall

Many studies of L_p, P_d and σ in the capillaries of frog mesentery and of amphibian and mammalian muscle report high values of σ for albumin, which do not accord with their rather high values of L_p. A capillary wall that could so effectively hold back albumin ought not, it seems, to be so permeable to water, if the two utilized the same pathway. Rasio (1977) reported similar findings for the retial capillaries of the eel, and high

Figure 14. Fluid exchanges across a capillary in a fish. *A*, Some possible values of hydrostatic and colloid osmotic pressure. *B*, The net inflow and outflow pressure if the capillary wall were impermeable to plasma protein.

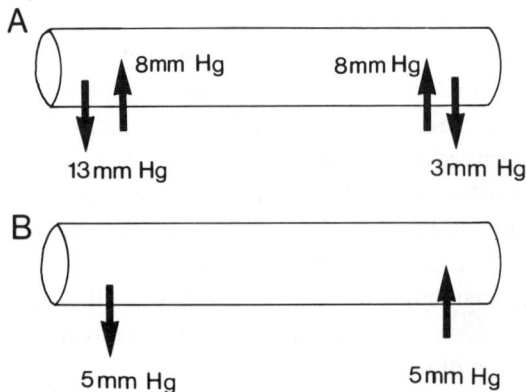

values for the permeability of tritiated water in eel retial vessels have also been reported by Stray-Pedersen and Steen (1975). Such studies suggested that water, small lipophilic molecules and ions such as Na^+ and K^+ move through pathways which are distributed over a greater area of the capillary than those which the macromolecules traverse.

Rasio *et al.* (1977) suggested that the entire inner surface of the capillary wall is available for the passage of water and small lipophilic molecules, whereas macromolecules exit via the clefts between endothelial cells, i.e. a paracellular pathway. We noted, earlier in this Chapter, the complex structure of these clefts, revealed in ultrathin sections (Bundegaard 1987). Perfusion studies show that a variety of macromolecules such as albumin, myoglobin and dextran can pass through them. In different capillaries, the σ for these macromolecules is unrelated to the magnitude of L_p; moreover, some of the highest values of σ are found in fenestrated capillaries, suggesting again that their passage across the capillary wall is not through the membrane of the endothelial cells but by some other route, i.e. the intercellular clefts. The permeability of these to macromolecules, is greatly increased by substances which chelate Ca^{2+} such as EDTA, but even 120% of EDTA, relative to the divalent ions present, leaves the permeability of eel retial capillaries to water unchanged (Stray-Pedersen 1975).

The effect of plasma proteins on permeability

The mismatch, mentioned above, between the high values of L_p and σ has led students of microcirculation to look for some further barrier which might hinder the movement of macromolecules. Myhre and Steen (1977), in a study of eel retial capillaries, demonstrated the great importance of plasma proteins in lowering capillary permeability both to water and, even more, to macromolecules. Perfused amphibian mesenteric capillaries in which the perfusate lacked plasma protein have elevated values of L_p (Michel 1983); the addition of cationized ferritin reduces it in a manner similar to albumin, and the iron could be visualized with the electron microscope, bound to the endothelial surface. It has been suggested that the molecules of ferritin act by changing the orientation of fibrous protein molecules at the endothelial surface, from a random array to an ordered matrix. This would reduce the number of randomly placed larger gaps between fibrous elements; the effect, it is believed, extends over the entire wall including fenestrations and intercellular clefts. In the capillaries of frog mesentery, between two thirds and four fifths of the resistance to fluid movement through their walls depends upon the interaction of the vessel wall with components of the plasma proteins.

Our functional picture of a capillary wall can thus now be integrated with our knowledge of its fine structure with some assurance. In brief, water and small lipophilic molecules pass out via a transcellular route and macromolecules exit through a paracellular pathway, the intercellular clefts; restricting and modulating both these is a lining layer in the form of an ordered fibre lattice, imposed by components of the plasma proteins on the endothelial membranes.

The fate of spilt plasma protein

How, may we ask, do fish avoid oedema, if plasma protein escapes steadily into the tissues? In higher vertebrates spilt protein is returned to the venous system by the lymphatic vessels. These are vessels of capillary dimensions composed of endothelial cells; they originate as blindly ending tubes amongst the tissues, which coalesce to form larger valved ducts that open into the venous system. There is an extensive but confusing literature discussing whether such structures occur in fish (Kampmeier 1969). Vogel (1985a) concludes that fish lack lymphatics comparable to those of mammals; they do not appear in the evolutionary sequence until the amphibians. Other, more localized, microcirculations may exist in the central nervous system and gut of fish, but our understanding of these is as yet incomplete.

Perhaps spilt protein is returned in the microvessels of the secondary blood system (Chapter 6). The narrow, coiled, anastomotic connections with the secondary arteries must impose a high resistance and secondary capillaries may well have a pressure so low as to favour uptake. However, the limited distribution of secondary capillaries, and particularly their absence from skeletal muscles, seem to make them a poor candidate for this role.

The capillary endothelium is, currently, an area of very active research. Cultured endothelium from pig vessels shows an asymmetric transfer of albumin, for the rate of transfer is greater into the lumen than out of it (Shasby and Roberts 1987), and it differs in endothelia of vessels from different sites. It is greater for the endothelium of pulmonary veins than for that of pulmonary arteries (Vecchio *et al.* 1987). Perhaps vessels at the venous end of the capillary bed in fish transfer protein into the capillary more readily than those at their proximal end, thus favouring its return. Other mechanisms are possible; there has to be some pathway for split plasma protein to re-enter the blood system. The problem cannot at present be resolved.

Veins

Veins provide wide-diameter channels of low resistance. By the time the blood in a fish has passed through the capillaries, very little of the pressure imparted to it by the heart is left. This pressure, the *vis a tergo*, is supplemented by a variety of venous pumping mechanisms which exploit the movement of adjacent organs and tissues. Where veins receive little support from adjacent tissues, as happens in the intestine, their walls have the same three layers as do those of arteries but these are thinner and have little elastic tissue in them (Figure 15C). Thus they are easily dilated to their full size by the small pressures available. The central ends of the anterior and posterior cardinal veins, and the veins of the liver, gonads and fins, are expanded to form large sinuses. They are strategically placed reservoirs of blood, available to fill the atrium as it is enlarged by cardiac suction each time the ventricle contracts. When veins run within resistant tissues their walls may be very thin. Many of the longitudinal veins of the trunk, such as the lateral abdominal vein and the cutaneous veins, have

Figure 15. Veins and their valves. *A*, A deep vein of the lower leg of man with parietal and ostial valves. *B*, The caudal vein of a dogfish, with only ostial valves. *C*, An ostial valve in the caudal vein viewed from the lumen. *A* redrawn from Dodd and Cockett (1976).

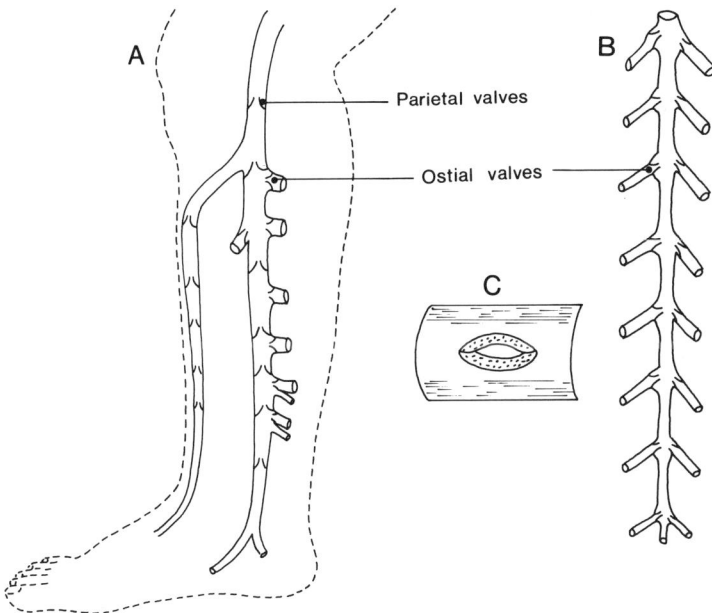

no muscle in their walls and may be little more than endothelium-lined channels in the dense connective tissue of the body.

Valves in veins

In Chapter 1 the point was made that blood has weight. In land animals the long columns of blood extending, for example, from the right atrium down to the feet, have the potential to increase pressure in the vascular beds of the lower limbs and upset the Starling balance across the capillary walls by favouring outflow and causing oedema. This hazard is lessened by valves which extend across the long veins of the limbs (Figure 15*A*) and, as these valves close and open in response to movements of skeletal muscles adjacent to them, they in effect segment the column of blood into a series of shorter lengths. Franklin (1937) termed these parietal valves. They occur only in veins subject to pressure from without, and they appear in the foetus at the time when it begins to execute movements of its skeletal muscles. The valves of the heart have developed before this.

Vertebrate veins bear a second type of valve, located in the openings of tributary veins into longitudinal veins; these are termed ostial valves (Figure 15*A*, *B*). Localized movements of skeletal muscles, in the area drained by the vascular bed of the tributary vessel, compress it and force the blood past the ostial valve (Figure 15*C*) into the main vein. Early students of the fish vascular system declared that fish veins lacked valves. They were wrong, but it is certainly true that fish veins lack parietal valves. There are no valves which can close across the length of such long veins as the lateral cutaneous, the lateral abdominal, the posterior cardinal, and the caudal vein. In a swimming fish such veins act as single conduits; they are not functionally segmented into a series of chambers, in which blood flows from one section to the next in a stop–start manner, as occurs in the deep leg veins of a man when walking.

The fish venous system is, however, richly endowed with ostial valves and these enforce a one-way flow from segmental regional vascular beds into the main longitudinal veins. Ostial valves are part of a more ancient circulatory system; they are located so as to exploit a feature retained more completely in primitive vertebrates, there metameric segmentation. When neuronal activity passes, as a chain reflex down the nerve cord, so do the myotomal muscles of the trunk contract in sequence. This forces blood past the ostial valves, at whatever level they occur, into the longitudinal vessel; once in the caudal vein or posterior cardinal vein, blood passes forward to the heart. Ostial valves are important components in a variety of auxiliary venous pumps, which assist the return of blood to the heart and which form the subject matter of Chapter 8. Parietal valves, in

Table 4. *Blood pressure in fish*

Species	Ventral aorta (mm Hg)	Dorsal aorta (mm Hg)	Temp (°C)	Authors
Atlantic hagfish *Myxine glutinosa*	6.6 ± 1.3	5.2 ± 1.3	7	Satchell 1986
New Zealand hagfish *Eptatretus cirrhatus*	10.8 ± 0.7	8 ± 0.8	14	Forster *et al.* 1988
Spotted dogfish *Scyliorhinus canicula*	31 ± 1	25 ± 1	13.5	Short *et al.* 1977
Trout *Salmo gairdneri*	39 ± 3.6	31 ± 3.8	10	Kiceniuk & Jones 1977
Lingcod *Ophiodon elongatus*	40 ± 0.3	29.3 ± 0.2	11	Farrell 1981
Cod *Gadus morhua*	46.4 ± 3.5	29.4 ± 2.4	13	Wahlqvist & Nilsson 1977

contrast, arise in response to the need to protect capillary beds from the disturbing effects of gravity. They first appear, in the evolutionary series, in the limb veins of the anuran amphibia (Suchard 1907). It is not surprising that they are absent from fish, buoyed up as they are by the water around them.

Blood pressure in fish
In teleost fish measurements of blood pressure have commonly been made in either the ventral or the dorsal aorta. The resistance of the gill circulation lies between these two sites and a quarter to a half of the pressure is lost across this.

In general it is true (Table 4) that there has been an increase in blood pressure in the evolutionary sequence and more active species have higher blood pressures than sedentary species. It increases during exercise and hypoxia but, as we shall see in Chapter 9, we still lack any real knowledge of how fish regulate their blood pressure.

Reactive hyperaemia
It has long been known that when a tissue becomes active or when its blood supply is briefly occluded, blood flow to it increases. This occurs in tissues in which the autonomic nerves have been cut, or their junctions

blocked, as it does in tissues which lack an autonomic supply. We noted above the existence of stress-sensitive ion channels in the endothelial cells, and the hyperpolarizing potential these produce when flow initially increases. This may spread via gap junctions which are known to couple them to the surrounding smooth muscle layer and relax this (Olesen *et al.* 1988). In addition, metabolites such as H^+ and lactate, and perhaps lowered oxygen tension, dilate arterioles and capillaries as do the endothelium-derived substances such as EDRF, prostacyclin and prostaglandins. Brief occlusions of the coronary arteries of mammals cause the release of adenosine, as well as EDRF (Nichols *et al.* 1988). Reactive hyperaemia is an intrinsic response and represents a primitive regulatory mechanism which tends to match the demands of a tissue for oxygen and nutrients to the supply of these by the blood. We will refer to it again in Chapter 10 when we examine the response of the circulatory system to exercise.

Figure 16. The principal blood vessels of the gills and head of a teleost fish. A. atrium, A.B.A. 1–4, afferent branchial arteries 1–4, A.C.S. anterior cardinal sinus or vein, B.A. bulbus arteriosus, D.A. dorsal aorta, D.C. ductus Cuvieri, E.B.A 1–4 efferent branchial arteries 1–4, H.V. hepatic vein, I.J.S. inferior jugular sinus, M.A. mesenteric artery to intestine, P.C.S. posterior cardinal sinus or vein, S.V. sinus venosus, V. ventricle, V.A. ventral aorta.

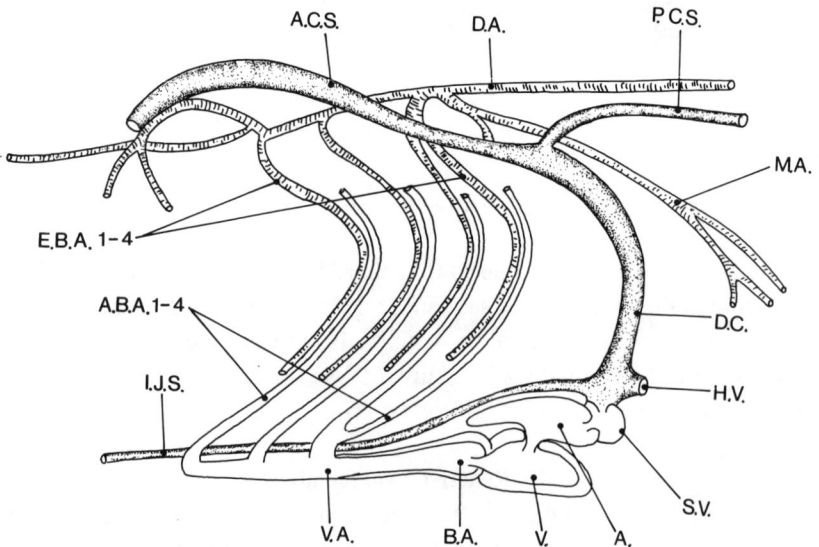

The principal blood vessels of the fish circulatory system

Some familiarity with the names of the principal blood vessels is necessary in order to discuss the integrative aspects of the circulation (Chapters 9, 10 and 11). The major vessels of fish are remarkably constant in position. The heart pumps blood into the ventral aorta and in most teleosts this divides into four pairs of afferent branchial arteries (Figure 16). From each of these, many filamentar arteries carry blood down the gill filaments to the arterioles that lead to the secondary lamellae, the afferent lamellar arterioles (Figure 19, below). Oxygenated blood is collected up by efferent filamentar arteries into four efferent branchial arteries which unite to form the dorsal aorta. This passes forwards beneath the spine, into the head, and backwards through the abdomen and into the haemal canal of the post abdominal trunk. In the head it divides into branches to the jaws, eyes and brain. In the trunk it gives rise to paired segmental arteries to the myotomes, fins and kidneys, and, in the abdominal region (Figure 16), to median arteries to the stomach, liver, intestine and gonads.

We have already noted that many of the veins give rise to large sinuses

Figure 17. The principal blood vesssels of the anterior part of a teleost fish. B. bladder, Br.H.P.V. branches of hepatic portal vein, B.V. bladder vein, C.V. caudal vein, D.A. dorsal aorta, G. gonad, H.A. haemal arch, H.P.V. hepatic portal vein, H.V. hepatic vein, K. kidney, L.L.L. left lobe of liver, P.C.S. posterior cardinal sinus, R.P.V. renal portal vein. Figures 16 and 17 redrawn after Allen's (1905) account of the blood system of the lingcod.

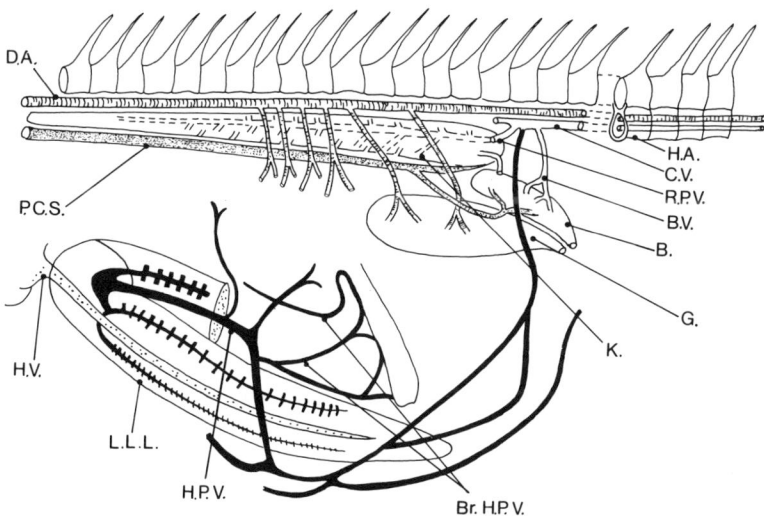

(Figure 16). The anterior cardinal sinus lie above the gills and receives blood from the branchial veins, and from the eye. A smaller ventral vessel, the inferior jugular sinus, drains the floor of the mouth. The posterior cardinal vein runs down each side of the midline on the dorsal wall of the abdominal cavity. At the caudal end of the abdominal cavity (Figure 17), the cardinal veins lie on and partly within the kidneys and are joined across the midline. They receive blood from the kidneys and the myotomes of the trunk. In elasmobranch fish a large vein, the lateral abdominal, encircles the abdomen just beneath the peritoneal epithelium lining the body wall. It is a ventral vein which links together the veins draining the pectoral and pelvic fins, and enters the ductus cuvieri on each side.

All of the blood from the post abdominal trunk, which in some fish may comprise half its length and a third of its weight, is returned in a single medial vessel, the caudal vein (Figures 17, 24, 25), which lies within the haemal canal. Rigid extensions of the vertebrae, termed the haemal arches, enclose and support it; this canal has an important circulatory function. It insulates the caudal vein from the backwardly moving waves of pressure generated by the myotomes during swimming. We shall consider this region further in Chapter 8.

All of the blood from the intestine and stomach (Figure 17) is collected up into veins which coalesce to form the hepatic portal vein. This enters the liver and subdivides to supply many thousands of sinusoids. The hepatocytes of the liver convert amino acids into peptides and proteins used in other parts of the body, a particularly important function in many fish as nearly 50% of their diet is protein. From the liver, blood is collected up into the hepatic vein, which is very short and enters the sinus venosus by paired openings. These lie opposite the entrance to the atrium and, as we noted in Chapter 1, are guarded in elasmobranchs by muscular sphincters which can regulate the outflow of blood from the visceral to the somatic circulation.

Mention must be made of a second portal system. At the point where the caudal vein leaves the haemal canal and enters the abdomen (Figures 17, 26), it divides into paired renal portal veins. Each of these subdivides to form capillary networks around the kidney tubules. Blood from these capillaries is collected up and enters the adjacent posterior cardinal veins. We will discuss the renal portal system at greater length in Chapter 6.

Those wishing to inject substances into live fish will find that the posterior cardinal vein is most often the vessel of choice. Its position, relative to the spine above it, should be ascertained in a cross section made immediately behind the heart of a dead specimen. It may then be easily entered from the ventral surface of a lightly anaesthetized specimen, if the needle is advanced until it just touches the spine.

Before we proceed in Chapter 6, to more specialized vascular beds we need to consider the blood and its abilities as a transport medium.

4

THE BLOOD

Blood is a suspension of differentiated cells in a specialized extracellular fluid, the plasma. The most abundant cells are the erythrocytes, or red blood cells, and the haemoglobin which they contain transports oxygen from the gills to the tissues. The protein moiety of the haemoglobin has also a crucial role in the transport of carbon dioxide, for its cations buffer the protons formed from the dissociation of carbonic acid; there is a complex and changing interrelation between the CO_2 in the plasma and the red cell as it passes from the tissues to the gills. The several kinds of leucocytes protect and repair tissues and are transported to sites where their services are needed. The neutrophils are phagocytic, and ingest bacteria. Monocytes move around the body in the blood and differentiate into tissue macrophages. Lymphocytes are carried in the blood to sites of injury and pathogen invasion, and there differentiate into plasma cells which produce specific antibodies. Other lymhocytes react directly to foreign invading cells. Plasma also contains specific proteins which bind and transport particular ions such as copper, iron and iodine. The anions of strong acids, such as chloride and phosphate, in solution in the plasma, are buffered by the presence of bicarbonate ion, and thus are prevented from exerting a large change of pH. This list does not exhaust the transportive functions of the blood; to do that other protective agents, nutrients, hormones, trace elements and vitamins would need to be included.

The crucial role of red blood cells and their contained haemoglobin can be judged by considering the problem of transporting oxygen in a 1 kg trout. At 10 °C its oxygen uptake might be 0.41 ml min^{-1} (Kiceniuk and Jones 1977). If its blood system contained only plasma, with a dissolved oxygen content of 0.3 ml dl^{-1}, it would require a cardiac output of 137 ml min^{-1} to transport the oxygen to the tissues, supposing they could extract all of it. The actual output is known to be 36.7 ml min^{-1}, i.e. one quarter of this, and the calculated value is impossibly high for a resting 1 kg trout. The difference reflects the amount of oxygen taken from the haemoglobin in the circulating red cells as they pass through the capillaries of the resting fish. Clearly, the amount of oxygen which can be carried by the blood when the haemoglobin is fully saturated, i.e. its oxygen capacity, is of great importance in determining how much can be transported to the tissues and how active the fish can be.

Haemoglobin

Haemoglobin is a metalloporphyrin, i.e. a combination of a haem unit which is an iron porphyrin, and globin which is a protein. The haemoglobins of different fish have identical haem units; it is the protein moieties which differ from species to species, and impart to them their particular properties. The atom of ferrous iron in the haem can associate with one molecule of oxygen; this is termed oxygenation. The oxygen is bound loosely so that the haem remains in the ferrous state, an ability imparted to it by the amino acids of the polypeptide chain. The haemoglobins of hagfish and lamprey consist, when oxygenated, of single chains, as does the myoglobin of muscles. Haemoglobins of elasmobranchs and teleosts are tetrameric, and consist of four chains. The four are not all alike, and commonly consist of two α and two β chains. The single-chain haemoglobins of hagfish have molecular weights of 16 500–17 000, and tetrameric haemoglobins are four times this, i.e. 66 000–68 000. Haemoglobin comprises as much as 70% of the dry contents of an erythrocyte, and its packaging within a cell membrane prevents its loss through the kidney of glomerular fish. In addition it prevents the haemoglobin from exerting the high osmotic pressure it would were it free in the plasma.

Blood oxygen capacity

Fish, like all vertebrates other than mammals, have nucleated erythrocytes which vary widely in shape and size from one species to another. Elasmobranchs and hagfish have particularly large cells; those of the rough-tailed stingray (*Dasyatis centroura*) measure 19.7 μm × 13.8 μm (Hartman and Lessler 1964). Many species have ellipsoid cells with diameters around 11–14 μm. The mean cell volume, MCV (measured in femtolitres, 1^{-15}) may differ by a factor of four if we contrast the large oval cells of the spiny dogfish with the smaller round ones of the tench (Table 5).

The abundance of red cells in the blood also varies greatly, and is

Table 5. *Haematocrits and haemoglobin in fish blood*

Species	Hct (%)	Haemoglobin (g dl^{-1})	MCHC (g %)	MCV (fl)	Ref.
New Zealand hagfish *Eptatretus cirrhatus*	12.6	3.0	24.2	538	1,
Spiny dogfish *Squalus acanthias*	11.5	2.9	18.8	650–1010	2, 3
Red paradise fish *Macropodus opercularis*	15.2	4.7	23	746	4
Tench *Tinca tinca*	24.1	6.8	33	245	5
Atlantic salmon *Salmo salar*	47	9.6	20.9	485	6
Nototheniid *Dissostichus mawsoni*	27	4.3	16	—	7
Nototheniid *Pagothenia bernacchii*	21	2.5	14.3	—	7

References: 1. Wells and Forster (1989); 2. Wells and Weber (1983); 3. Larsson *et al.* (1976); 4. Weinberg *et al.* (1973); 5. Eddy (1973); 6. Sandnes *et al.* 1988; 7. Wells *et al.* (1980).

indicated by the haematocrit (Hct), i.e. the volume of red cells, expressed as a percentage, of a centrifuged sample of blood. Table 5 might well have had the icefish (*Chaenocephalus aceratus*) as its first entry, as this species has scarcely any erythrocytes (Ruud 1954) and these lack haemoglobin altogether. Most fish have Hcts less than those of mammals, but fast-swimming predators like blue marlin (*Makaira nigricans*), with an Hct of 43% and mackerel, 52.5%, span the value of 47% in man. Even amongst the antarctic nototheniids, *Dissostichus mawsoni*, a midwater predator, has a notably higher Hct than *Pagothenia bernacchii*, a sedentary benthic scavenger (Wells *et al.* 1980). Haemoglobin content ranges from the low values of 2–3 g dl^{-1} (Table 5) of those antarctic fish that have haemoglobin, up to the levels reported for yellowfin tuna (*Thunnus albacares*), 15.8–18.9 g dl^{-1} and frigate mackerel (*Auxis rochei*), 17.8–21.2 g dl^{-1}, which rival that of human blood (13–18 g dl^{-1}) (Klawe *et al.* 1963).

The Hct and the haemoglobin content of the blood tend to be correlated; with fewer red cells there is less haemoglobin in the blood. However, mean corpuscular haemoglobin content (MCHC) is by no means the same in different species. MCHC is measured in g 100 ml^{-1} of packed red cells. The MCHC of man ranges between 32 and 38 g %. In slow-moving and sedentary species such as plaice and angler fish it tends to be less than in active fish like herring and mackerel. The blood of antarctic nototheniids has a low haemoglobin content, not only because it has fewer blood cells in it but also because each cell has a low MCHC (Table 5). Blood volumes and haemoglobin values tend to be inversely correlated; hagfish have large blood volumes, around 18% compared with teleost values of 5–6%, but the blood has a low haematocrit and a low Hb content.

The oxygenation of haemoglobin

Haemoglobin has the ability to combine reversibly with oxygen which is loaded on when the pressure is high enough and is unloaded when it falls. The reversibility of oxygenation is a property imparted to the haem by its association with the polypeptide chain, and involves a configurational change in the haemoglobin molecule. The binding of oxygen to the haem causes the two β chains to move closer together so that the distance between the two iron atoms is reduced. Such site–site interactions are the basis of cooperativity; they cause the affinity of the second haem for oxygen to be greater once the first haem has been oxygenated.

When the partial pressure of oxygen in the blood is plotted against the amount of oxygen bound to the haemoglobin, the oxygen dissociation curve is derived, and such curves have been determined for many species of fish (Figure 18), and are always non-linear. Cooperative interactions result in a sigmoidal oxygen dissociation curve. In the absence of such interactions, and particularly in monomeric haemoglobins, the curve is hyperbolic, as is expected from the simple mass law relationship.

Whereas cooperative interactions increase the affinity of the particular haemoglobin for oxygen, and are an example of intramolecular or homotropic reactions, other, heterotropic reactions decrease affinity. The presence of protons, CO_2, various organic phosphates, and increasing temperature reduce the amount of oxygen bound at a particular partial pressure.

Hill (1910) developed an equation on the supposition that binding oxygen depends on subunit interaction.

$$Y = \frac{Kp^n}{(100 + Kp^n)}$$

The blood

Y = % of Hb combined with O_2, p = partial pressure of O_2, K = the equilibrium constant, n = a constant, for a particular Hb, of the degree of interaction

If $n = 1$, there is no cooperative interaction and the dissociation curve is hyperbolic (Figure 18A). The larger the value of n, the more evident is the sigmoidicity of the curve; Figure 18B shows the dissociation curve of the antarctic nototheniid *Dissostichus mawsoni* which, living in waters close to freezing, is always subject to high tensions of oxygen.

Oxygen affinity, P_{50} and environmental P_{O_2}

The oxygen affinity of haemoglobin is conveniently indicated by the P_{50},

Figure 18. Four factors which can alter haemoglobin oxygen affinity. A, CO_2. Equilibrium curves for whole blood of hagfish *Eptatretus cirrhatus* at pH 7.8, 16 °C. B, pH. Blood of an antarctic fish, *Dissostichus mawsoni* at -1.5 °C. C, Temperature. Trout blood at $P_{CO_2} = 1.0$ mm Hg. D, Nucleoside triphosphates. Whole blood from hypoxic and normoxic carp; inset shows levels of their nucleoside triphosphates (NTP). Redrawn, A from Wells *et al.* 1986, B from Tetens and Wells 1984, C from Eddy, 1971, D from Weber and Lykkeboe 1978.

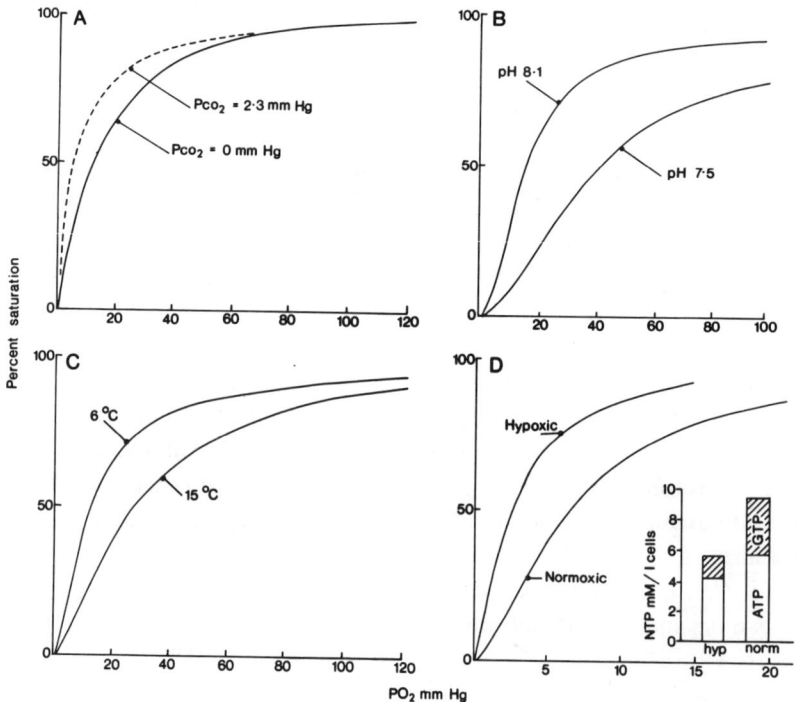

i.e. the partial pressure of oxygen at which half the haemoglobin is saturated. The P_{50} of the bloods of fish that live in well-aerated waters tends to be higher than those that live in still waters, e.g. at 10 °C and a P_{CO_2} of 0.3 mm Hg, the blood of trout has a P_{50} of 14 mm Hg (Figure 18C), and carp 5–7 mm Hg (Figure 18D).

The oxygen dissociation curves of fish from waters with a high oxygen tension tend to lie to the right of those from ponds and less well-aerated waters. Powers (1980) compared the oxygen dissociation curves of cata-stomid fish which were lotic water breathers, i.e. those that live in the slowly flowing sides of streams where oxygen is less abundant, with those that frequent the swiftly flowing centre. The two sets of curves were quite distinct and above a P_{O_2} of 13 mm Hg they did not even overlap. Slow zone species have bloods with higher oxygen affinities and can live in water with a lower oxygen content. Some 50 or so, largely unrelated, species of fish have evolved the ability to breathe air and they use a variety of accessory respiratory organs to do this. Air is a much richer source of oxygen than water and such species have bloods with oxygen dissociation curves to the right of those of obligate water breathers. One of these, the African lungfish, aestivates during periods of drought; it remains buried below the surface of the mud in a cocoon and its metabolic rate falls to only 10% of that of the awake fish. The dissociation curve is shifted to the left, and the P_{50} falls from 33 to 9 mm Hg at a pH of 7.5 during aestivation (Johansen *et al.* 1976).

A fish with a blood of relatively high P_{50}, such as trout, can still load up its haemoglobin at the gills because of the high partial pressure of oxygen there, and has the advantage that the tissues do not have to endure a very low pressure of oxygen in order for the bound oxygen to off-load and be made available. Fish that live in waters of low oxygen partial pressure, such as carp, may, if they have a blood with a low P_{50}, still be able to load it up fully at the gills, because of its high affinity for oxygen, but must endure the disadvantage that, as the blood circulates through the tissue capillaries, it will not off-load its oxygen until the partial pressure in the tissue fluids has fallen to a low level. But such fish will be better able to survive hypoxia for, despite the low environmental P_{O_2}, the blood leaves the gills more fully saturated. The spiny dogfish is viviparous and the blood of the 'pups', which live for a time prior to birth within the uterus, has a P_{50} of 11.5 mm Hg compared with the 13.2 mm Hg of the mother (Wells and Weber 1983). This may reflect the fact that the intrauterine water is oxygenated from the maternal blood stream, and there has to be a gradient of pressure for this to occur.

The blood

Table 6. *Blood parameters in fish*

Species	Temp. (°C)	P_{50} at P_{CO_2} (mm Hg) or pH	Bohr factor	n	O_2 cap. (vols %)	Ref.
Kawakawa *Euthynnus affinis*	25	41.5, pH = 7.13 10, pH = 7.95	− 0.59	1.72	—	1
Nototheniid *Pagothenia borchgrevinki*	− 1.5	31.1, pH = 8	− 0.26	1.92	7.7	2
Spiny dogfish *Squalus acanthias*	15	13.2, P_{CO_2} = 2.2	− 0.28	1.65	3.5–4.5	3
Hagfish *Eptatretus cirrhatus*	15	12.3, pH = 7.8	− 0.43	1.38	2.3	4
Trout *Salmo gairdneri*	6 15 15	8.7, P_{CO_2} = 0.3 15.8, P_{CO_2} = 1.0 34, P_{CO_2} = 6.0	− 0.54 − 0.57	— —	10.4 8.9	5
Burbot *Lota lota*	15	9.3, P_{CO_2} = < 0.3	—	2.3	6.5	6
Tench *Tinca tinca*	5 13	2, P_{CO_2} = 2.5 3, P_{CO_2} = 2.5	− 0.75 − 0.64	0.8–1.7	8	7

References: 1. Jones *et al.* (1986); 2. Tetens and Wells (1984); 3. Wells and Weber (1983); 4. Wells *et al.* (1986); 5. Eddy (1971); 6. Cameron (1973); 7. Eddy (1974).

The Bohr effect

We noted above that increasing the concentration of protons decreases the affinity of many haemoglobins for oxygen, a phenomenon known as the Bohr effect. It can be defined as the change in the P_{50} with unit change of pH, and quantitatively by the expression, $\Delta \log P_{50}/\Delta \log pH$. Fish which live in well-aerated waters tend to have a large Bohr effect. That of kawakawa at 25 °C is − 0.59 (Jones *et al.* 1986), of trout at 15 °C is − 0.57 (Eddy 1971), and of tench at 13 °C is − 0.64 (Eddy 1973) (Table 6). Carbon dioxide dissolves in water to form carbonic acid which ionizes to yield protons. In the gills the haemoglobin is subject to a very low P_{CO_2} because this gas is very soluble and diffuses rapidly into the surrounding water. As the blood passes into the tissues it encounters a higher P_{CO_2} and bloods with a significant Bohr factor will off-load some of their oxygen without the necessity for a fall in the tissue P_{O_2}.

Conversely, when the blood returns to the gills, the loss of its CO_2 increases its affinity for oxygen and assists on-loading. The extent of the vertical displacement of two oxygen dissociation curves plotted at arterial and venous pH indicates how much oxygen will be off-loaded purely as a result of the Bohr effect (Figure 18*B*). If the blood of a trout, in passing from the gills to the tissues, encounters a pH change from 7.9 to 7.5, it will release one third of its oxygen without any fall in P_{O_2} (Eddy *et al.* 1977). In the antarctic nototheniid *Dissostichus mawsoni* (Figure 18*B*), a change of pH from 8.1 to 7.5 would release 40% of the bound oxygen (Tetens and Wells 1984). Wells (1989) remarks that the Bohr shift 'exploits an arterio-venous pH difference to provide a feed-back control of oxygen release for carbon dioxide gain'.

It needs to be realized that the actual oxygen dissociation curve which blood undergoes as it passes round the circulation is compounded of these two curves; it starts from the gills along the arterial curve, and gradually changes to the venous curve as it passes through the capillaries and into tissue environments of increasing proton concentration.

The Root effect

When first described, the Root effect (Root 1931) was regarded as quite distinct from the Bohr effect. The latter term applied to the phenomenon outlined above, that an increase in the concentration of protons lowered the affinity of the haemoglobin for oxygen. Acidification of the blood of fish possessing a Root effect haemoglobin reduced, it was believed, the capacity of the blood to hold oxygen at any pressure, no matter how high. Scholander and Van Dam (1954) described haemoglobins which, when acidified, could not be saturated at oxygen pressures of 100 atm. Brittain (1987) lists 83 species of fish known to possess Root effect haemoglobins; that of the Dover sole (*Microstomus pacificus*) is like many others. At a P_{O_2} of 150 mm Hg it is only 45% saturated if the pH is 5.5 (Ingermann and Terwilliger 1982). The haemoglobins of trout, carp, goldfish, eel, tuna and menhaden (*Brevoortia tyrannus*) all show a Root effect and have been subject to detailed studies of recent years.

The Root effect is now regarded as an extreme development of the Bohr effect (Brittain 1987), but such haemoglobins do show features of structure and reactivity that are peculiar to them. Root effect haemoglobins show not only a great reduction of oxygen affinity on acidification, but also a loss of cooperativity such that *n* may fall below unity. This is not a feature of the Bohr effect. These changes are caused by interaction between the α and β chains; Pankhurst and Goss (1984), by forming hybrids between the β chain of carp and the α chain of man, the so called 'mermaid' haemoglobins, have identified certain amino acids in the chains which are crucially important if these attachments are to occur. One particularly, the serine residue at F9 on the β chain, has to be present if the haemoglobin is to show a Root effect, although not all haemoglobins with it necessarily do so. These interchain reactions profoundly depress the affinity and cooperativity when pH is lowered, but this does not imply that more oxygen could not be on-loaded if the pressure were yet further raised. The concept that there is some haemoglobin disabled by protons, from taking up oxygen under any pressure, is now doubted.

Three sections hence we will consider the effect of certain organic phosphates on haemoglobin oxygen affinity, but we can note here that the Root effect is often powerfully influenced by them. Raising their concentration augments the Root effect and increases the proportion of haemoglobin unsaturated at a particular P_{O_2}.

The Root effect haemoglobins play a significant part in concentrating oxygen in the ocular fluids and in the swimbladder of many teleost fish. By acidifying the blood after it has passed through the arterial vessels of a rete, and prior to is entry into the venous vessels, the haemoglobin is caused to unload some of its oxygen from combination into solution and the P_{O_2} is raised. This leads either to a high P_{O_2} in the ocular fluids, or to the liberation of gaseous oxygen into the swimbladder. These topics will be discussed in greater detail in Chapter 7.

The effect of cell haemoglobin concentration

Lykkeboe and Weber (1978) demonstrated that when solutions of carp haemoglobin are diluted, their oxygen affinity increases because dilution dissociates complexes with cofactors such as the erythrocytic phosphates, to be considered shortly. Hypoxia leads to swelling of the erythrocytes (Chiocchia and Motais 1989) and a consequent dilution of the haemoglobin within. Soivio *et al.* (1974) reported that the swollen erythrocytes of asphyxic trout, incubated with oxygen, diminished in size, and after 160 min the MCHC (g l^{-1}) had risen from 165 to 182, indicating that the volume of the swollen cells had decreased. We will consider this topic again in Chapter 11.

The effect of temperature

Increasing temperatures reduces the oxygen affinity of fish blood as can be seen for the bloods of trout and tench in Table 6. The effect is related to the Bohr shift because P_{CO_2} and pH are related to temperature. As temperature is raised the solubility of CO_2 in blood and plasma decreases, and the ionization of non-bicarbonate buffers, such as proteins, increases, as does the ion product of water (Eddy 1971). At 25 °C pure water is neutral and both its pH and its pOH are 7.0. As temperature falls its pH has to rise, if neutrality is to be maintained. Thus as temperature increases pH falls, so that the ratio of hydrogen to hydroxyl ions is maintained and the oxygen affinity of the blood increases. Trout blood (Figure 18C), at 6 °C (P_{O_2} = 100 mm Hg, P_{CO_2} = 1 mm Hg), has a pH of 8.13, and it is 92% saturated: at 20 °C its pH is 7.6 and it is 88% saturated (Eddy 1971). The blood of tench is known to behave similarly (Eddy 1973), although at the same temperature, the pH and the $[OH^-]/[H^+]$ ratios are higher. Whereas in warm-blooded vertebrates this relationship can be exploited to advantage because active muscles have higher temperatures and the off-loading of oxygen is assisted in active tissues, in fish this is not usually so for most fish are poikilothermic. The temperature of their muscles is rarely more than 1–2 °C above that of the surrounding water. Indeed, high temperatures hinder oxygen uptake at the gills, and levels of hypoxia which can be survived in cool water may prove lethal at higher temperatures.

A remarkable exception to this relationship is seen in the blood of the albacore, which has a reversed temperature effect. Warming albacore blood increases its affinity for oxygen; the effect is most noticeable if the blood is warmed rapidly (Cech *et al.* 1984). The molecular mechanism involved is unknown. The authors suggest that it favours the operation of the retial heat exchangers in the pathway to the red muscle. The diffusion of heat from the warmed venous blood into the arterial blood flowing to the red muscle increases its oxygen affinity, lowers its P_{O_2} and diminishes the likelihood that the heat exchanger will short-circuit the muscle circulation by allowing oxygen to diffuse directly across the retial vessels into the venous blood. The topic will be discussed in greater detail in Chapter 7.

The effect of nucleoside triphosphates (NTP)

Fish, like other vertebrates, can regulate the affinity of their blood for oxygen by varying the amount of certain organic phosphates in the red cells. The ones most commonly reported in fish are 2,3-diphosphoglycerate (DPG), adenosine triphosphate (ATP), guanosine triphosphate (GTP), and inositol pentaphosphate (IPP), although others have been reported. DPG binds to the deoxygenated tetrameric haemoglobin forming salt bridges with seven of the amino acids of the β chains and this depresses the affinity of the haemoglobin for oxygen and shifts the oxygen dissociation curve to the right. Decreasing the level of NTP increases the oxygen affinity of the blood (Figure 18D) and lowers the P_{50} (Weber *et al.*

1983). Kono and Hashimoto (1977) have shown in trout, carp, and the yellow tail (*Seriola quinqueradiata*), that GTP was more effective, in equimolar solutions, than ATP. Acclimating the antarctic fish *Pagothenia borchgrevinki* from -1.5 to $+4.5\,°C$ increased the ATP expressed as mmol l^{-1} red cells from 0.7 to 1.8 and raised the P_{50} of the blood from 20.6 to 27.6 mm Hg.

The response is a relatively slow one. Soivio *et al.* (1980) report that the erythrocytic ATP of the blood of trout acclimated to water with a P_{O_2} of 37–52 mm Hg, fell from 4.45 to 2.51 μmol ml^{-1} and the oxygen affinity increased by 25%. This response was the slowest of a triad of responses which also included an increase in oxygen affinity due to the Bohr effect, and an increase in oxygen capacity due to the increase in Hct. The fall in erythrocytic ATP was not complete until a week had elapsed. Greaney and Powers (1977) found that the ATP level of the blood of the salt water minnow (*Fundulus heteroclitus*), subject to hypoxia, did not attain its lowest level of ATP until the 10th day. Tetens and Lykkeboe (1981) showed, in trout, that the response is a graded one; waters of P_{O_2} 150, 80 and 50 mm Hg resulted, after 12 days, in bloods (at pH 7.8 and 20 °C) with P_{50} values of 24.1, 21.7 and 16.8 mm Hg, respectively. The blood O_2 affinity was inversely correlated in stepwise manner with the level of erythrocytic ATP. Anoxic incubation of trout blood *in vitro* resulted in a rapid reduction in erythrocytic ATP, with a $t_{\frac{1}{2}}$ of 75 min, i.e. a much more rapid response than occurs *in vivo*. The authors suggest that this direct red cell response is itself regulated within the fish by some as yet unknown mediator.

In the antarctic fish *Pagothenia borchgrevinki* exposure to a P_{O_2} of 60 mm Hg for 11–14 days, at $-1.5\,°C$ leads to a fall in the P_{50} from 31 in normoxic fish to 21 mm Hg (Wells *et al.* 1989); the response is accompanied by a 27% fall in the concentration of erythrocytic ATP. The study is of special interest as the fish is one which is unlikely, in the Antarctic environment, ever to experience hypoxia. It suggests that the adjustment of oxygen affinity by changing the concentrations of erythrocytic NTP is not an adaptive mechanism in the sense that it has arisen through natural selection, but is rather a mechanism of acclimation concerned with adjustments within the individual, of a more ephemeral nature. The increase in the oxygen affinity of the blood of aestivating lungfish, noted above, is known to be due to a reduction in its level of GTP (Weber *et al.* 1977).

Acclimation and adaptation in haemoglobins
Wells (1989) has cautioned against facile adaptionist thinking when interpreting such features as oxygen dissociation curves and Bohr coeffi-

cients. Adaptation implies the accumulation of favourable gene mutations through the process of natural selection, and characteristics of the dissociation curves of fish bloods have often been cited as examples of this. Yet, as we shall see in Chapters 10 and 11, the responses of a fish to exercise and hypoxia are complex and depend on many features of the respiratory and circulatory systems which are integrated together. As Wells (1989) remarks, panselectionist physiologists tend to atomize animals into sets of traits or features, each of which is supposedly, separately optimized, by natural selection, for its particular role. It would indeed be surprising if the ability of a species of fish to respond appropriately to respiratory challenges depended only on the genetic determinants that specified the nature of its haemoglobin. The study by Cech *et al.* (1984) of the blood of the albacore tuna, a fast-swimming fish of the open ocean, shows it to possess a hyperbolic oxygen dissociation curve. Adaptionist thinking would have guessed it to be highly sigmoid and set to the right of most other species. In fact its P_{50} at 25 °C and $P_{CO_2} = 0$, is 8.3 mm Hg, below that of the spiny dogfish. But this is not true of other species of tuna; the dissociation curve of kawakawa is sigmoid, with a Hill's number of 1.7 and a P_{50} of 21 mm Hg at a P_{CO_2} of 5 mm Hg (Jones *et al.* 1986). In each species, we must presume, multiple constraints are optimized both by selection and by ephemeral processes such as acclimation, in one of many possible combinations, a number of which are potentially advantageous.

Multiple haemoglobins

Many fish are known to possess haemoglobins which differ in amino-acid composition, electrophoretic mobility, oxygen affinity and the magnitude of the Bohr effect; some salmonids exhibit a total of 18 electrophoretically different haemoglobins during their life cycle. Sharp (1973) studied the haemoglobins of 31 species of marine fish and found only three which had a single haemoglobin. The blood of coho salmon (*Oncorhynchus kisutch*) (Giles and Randall 1980), has 10 components; three of these (designated A6–A8) comprise 50–55% of adult blood but over 95% of fry blood. The different proportions give the two bloods different oxygen affinities. Fry blood has a P_{50} of 3.9 mm Hg at a pH of 8.5 and a temperature of 9.8 °C and it has a high Bohr shift. Adult blood has a P_{50} of 14 mm Hg at the same temperature and pH and it has a much smaller Bohr shift. The cathodal components of the haemoglobin of sockeye salmon (*Oncorhynchus nerka*), C2 and C2, have a higher affinity for oxygen than the anodal component A1; their respective values of P_{50} at 15 °C are 13.7, 14.4 and

27.9 mm Hg (Sauer and Harrington 1988). Oxygen binding by the anodal component is very sensitive to pH over the range 7.0–8.0; it has a Bohr factor of − 0.93 between pH 7.0 and 7.7, whereas the affinities of C2 and C3 are unaffected by acidity between pH 7.0 and 8.0. In Chinook salmon, it is known that the anodal components are present in the fry and throughout the entire life; additional cathodal components are added as the fry mature to become adult (Harrington 1986). Spawning and early development of most species of salmon occur in fresh water whereas maturation occurs largely in the ocean. Giles and Randall (1980) interpret this type of haemoglobin multiplicity as an adaptation to the high rate of oxygen consumption of the small fry; it enables them to use a larger proportion of the oxygen transported by the blood than do the adults. But it has to be said that in many other species of fish the multiple haemoglobins are very alike and adaptive explanations are lacking.

The carriage of CO_2 in the blood

Carbon dioxide from the tissues enters the blood, but little of it is transported in physical solution for it is reversibly bound, both in the red cells and in the plasma. In the red cell, CO_2 and water are rapidly catalysed to carbonic acid which dissociates into protons and HCO_3^- in solution. This catalysis results from the presence of a zinc containing enzyme, carbonic anhydrase, which is located both in the membrane and in the cytosol of the cell (Smeda and Houston 1979). The protons would acidify the blood were it not that haemoglobin, when deoxygenated, acts as a buffer and mops them up. The uncatalysed rate of the reaction is so low that, particularly at temperatures below 20 °C, very little bicarbonate would have formed by the time the blood had passed through the tissue capillaries. The speedy catalysis of the reaction by carbonic anhydrase is a prerequisite for the off-loading of oxygen in the tissues due to the Bohr effect. The bicarbonate ion, in turn, diffuses from the red cell and pairs with cations, chiefly Na^+, in the plasma; at the same time Cl^- moves into the red cell from the plasma. This exchange, the 'chloride shift' is a familiar feature of the blood of air-breathing vertebrates, and Cameron (1978) has confirmed that it occurs also in trout. The net effect of these exchanges is to increase the osmotic pressure of the red cell, and water enters, causing it to swell.

The buffering power of the blood of a fish is contributed partly by the bicarbonate system and partly by the titratable groups of the protein component of haemoglobin. In the physiological range between pH 6 and 9, buffering by Hb concerns the imidazole group of its histidine residues.

Cyclostome and teleost haemoglobins are characterized by low histidine content and low buffering values; they have 2–6 histidines per chain. Elasmobranchs have haemoglobins with high buffering values and resemble higher vertebrates with typically 8–12 groups per chain (Jenssen 1989). The amount of the bicarbonate buffer component tends to be related to the rate at which the species generates CO_2 and hence protons, and is high in elite performers such as the blue marlin. Buffering power is expressed in slykes, β, defined as mmol of base required to alter the pH of a homogenate of tissue, by one pH unit per gram of tissue, between pH 6 and 7; that of the blood of the blue marlin is 21.3 slykes (Dobson *et al.* 1986). This is five times that of the blood of the relatively inactive New Zealand hagfish, in which $\beta = 4$ slykes (Wells *et al.* 1986).

In the large central veins, where blood flows slowly, some bicarbonate is formed in the plasma at the uncatalysed rate; the plasma bicarbonate concentration increases and CO_2 partial pressure falls. But this is only a small proportion of that formed catalytically in the red cells.

Blood from the central veins, after passage through the heart, enters the secondary lamellae, where it is exposed, on one side of the very thin lamellar membrane, to the virtual CO_2 vacuum of the exterior water. The sequence of changes that occurred in the tissues now takes place in the reverse direction, and the CO_2, carried largely in the blood as bicarbonate, passes via the intermediary of the plasma, through the epithelia of the lamellae and into the water. Some carbonic anhydrase is located in their epithelial cells, but the part it plays is small compared with that in the red cells (Henry *et al.* 1988). There is some parallel between the carriage, by the blood, of oxygen and carbon dioxide. For both the red cell acts as a sink, and the gases pass through the intermediary of the plasma to it, but for carbon dioxide the tissues are the source whereas for oxygen the source is the gill epithelium.

The plasma proteins

Plasma contains an assemblage of proteins that differ widely in individual abundance, molecular weight and function. Collectively they contribute to the colloid osmotic pressure, the importance of which, in maintaining the Starling balance across the capillary endothelium was discussed in the previous Chapter. Others are separately concerned in transporting agents as various as copper, iron, iodine and lipids around the circulation.

Larsson *et al.* (1976), in a study of the plasma proteins of 27 marine fish from the Skagerrak, found their abundance ranged from 1.7 to 5.8 g dl^{-1}. Sulya *et al.* (1961), in an earlier study on 26 species from 14 families of fish from the Gulf of Mexico, found values of 1.7–6.2 g dl^{-1}. The highest values approach the 7.2 g dl^{-1} of human blood, but the mean value of the species studied by Larsson *et al.* (1976) was only 3.9 g dl^{-1}. There appears to have been an increase in plasma protein concentration in the course of evolution and the mean value of the six elasmobranch bloods investigated was 2.6 g dl^{-1} compared with 4.1 g dl^{-1} for the 19 species of teleost. We have already noted that blood pressures increase through the evolutionary series (Table 4). Plasma protein concentration in fish is less stable than in mammals, and the stress caused by handling depresses it for some days, as does starvation. Wide variation is found in the levels in Atlantic salmon which, as they do not feed in fresh water, metabolize their plasma proteins.

Electrophoretograms of fish plasma show many components, none of which corresponds completely to those of mammals. The plasma of trout has been extensively studied by Perrier *et al.* (1973, 1974, 1976, 1977) and in their initial study 13 bands were recognized; band 12 is fibrinogen, to be considered shortly in its role as one of the proteins responsible for blood clotting. It may be identified as the protein present in the plasma but absent from the serum derived from the clot. Bands 7 and 8 are ceruloplasmins, i.e. specific transport proteins which bind copper. Band 4 is transferrin, another transport protein which binds iron (Perrier *et al.* 1974). Band 1 contains an iodide-binding protein. It serves to facilitate iodide uptake at the gills and to limit its loss through the kidneys (Perrier *et al.* 1976). A component in bands 2 and 3 can bind haemoglobin, but unlike human haptoglobin, the binding is partial and reversible (Perrier *et al.* 1977). Of the 13 bands six have been identified as glycoproteins and two as lipoproteins. In Chapter 2 we noted the existence of plasma lipoproteins of high molecular weight (Smith *et al.* 1988); these are supramolecular complexes of lipids and polypeptides that transport dietary and endogenously synthesized lipids around the body. A hydrophobic core of triglyceride and cholesteryl ester is surrounded by a surface layer of phospholipid, cholesterol and a collection of apoproteins. In the channel catfish the major lipoprotein of the VLDL and LDL classes has a molecular weight of 250 000.

Whether fish plasma contains albumin has been much discussed and the term 'albumin-like' has been used by earlier authors with reference to the peak of more mobile components which are free of carbohydrate, at the anodic end of the electrophoretogram. In the plasma of trout, bands 2 and 3 are 'albumin-like' (Perrier *et al.* 1977). The major function of plasma albumin in mammals is to transport free fatty acids. Davidson *et al.*

(1989), using the criterion of ability to bind palmitate, identify as albumins two proteins in salmonid plasma, of molecular weight 68 000, that occur in bands 2 and 3. The authors note that these albumins have a lower mobiity than mammalian albumin and occur in the position normally occupied by an α globulin. Their amino-acid compositions, whilst broadly similar to mammalian albumin, differ in some respects. Their content of methionine is double that of man and they have 19 rather than the 35 cysteine residues that characterize mammalian plasma albumins.

Of still smaller molecular weight are the 'antifreeze' glycopeptides, so important in fish that dwell beneath the ice. In antarctic nototheniid fish, glycopeptides with molecular weights ranging from 34000 down to 10 500 depress the freezing point of the plasma from -1 °C down to -2.2 °C, in sea water with a freezing point of -1.9 °C (Ahlgren *et al.* 1988).

Important amongst the larger globulins are the circulating immuno-globulins which enable fish to mount a defence to particular, viral and bacterial pathogens. Two of the four immunoglobulins known for mammals have been recognized in elasmobranchs (Rosenshein *et al.* 1986). The largest is like the IgM found in man with a molecular weight around 900 000 and is believed to be pentameric. A smaller monomeric immunoglobulin corresponding to IgG, of molecular mass 180000 follows this in sequence, when the immune system responds to an antigen. Each monomer consists of a light and a heavy chain, as in higher vertebrates. Teleosts have a tetrameric immuoglobulin (Lobb 1986). Macrophages may play a part in processing antigens taken up by phagocytosis prior to passing them on to lymphocytes in which the immunoglobulins are synthesized (chapter 5). Fish have an active anti-body response and a competent immunological memory.

Fibrinogen and the blood clotting factors

Blood loss due to injury or parasite invasion is controlled by the for-mation of a clot and the mechanism of clotting in fish (Smit and Schoon-bee 1988) is, in essentials, like that in mammals (Allison 1989). A series of coagulation factors and of zymogens (proenzymes) each sequentially activates the next, leading to the formation of a clot of fibrin in which cells are enmeshed. This plugs the site of injury. The active forms of the zymogens are serine proteases each of which activates the next factor in the sequence by splitting from it a few peptide bonds to uncover the active enzyme site. The succession of these proteolytic reactions results in a cascade effect with amplification at each stage so that even a small initial reaction will result in an effective clot.

Smit and Schoonbee (1988) have identified in the plasma of tilapia (*Oreochromis mossambicus*), the following coagulation factors: I fibrinogen, II prothrombin, III tissue factor, IV calcium, V proaccelerin, VII proconvertin, VIII antihaemophilic factor, IX Christmas factor, X Stuart Power factor, XI plasma thromboplastin antecedent and XII Hageman factor. Human plasma contains 300 mg dl^{-1} of fibrinogen; that of tilapia contains 185 mg dl^{-1} which is almost within the lower limit, 190 mg dl^{-1}, of human values. Factors II, VII and IX–XII are proteases, the others are cofactors.

The traumatic coagulation time in tilapia is 22.7 sec which is about twice that of human blood. Coagulation time in glass tubes is 140.7 sec. The difference reflects the fact that there are two pathways, the intrinsic and the extrinsic, which lead up to factor X. The components of the intrinsic pathway are already present in the blood and are activated when they come into contact with tissue elements such as collagen fibres, which lie beneath the endothelium or pillar cells of the peripheral vessels and gill lamellar blood spaces. Such surfaces are likely to be exposed when tissues are injured. This phase of contact activation involves factors XI and XII. It in turn activates factor IX, and in the presence of Ca^{2+} and factor VII, leads to factor X. The extrinsic pathway, which operates more rapidly, involves the liberation of tissue factor, a cofactor from damaged tissue which, in the presence of Ca^{2+} activates in turn factor VII and factor X. It is this pathway which is activated in the common situation in which the epidermis of the scales is injured and the superficial capillaries are damaged; tissue factor and Ca^{2+}, from the sea water, are both then likely to be abundant.

In mammals the blood contains non-nucleated platelets which assist in haemostasis by initially aggregating and plugging gaps in the endothelium, and also play a part in the activation of factor X. In the presence of thrombin they liberate serotonin and potassium, which play a part in firming and contracting the clot. In fish the place of platelets is taken by nucleated leucocytes called thrombocytes, noted above, and thrombin helps agglutinate them to form a haemostatic plug (Lewis 1972).

Elasmobranchs appear to have clotting times of several hours when the blood is withdrawn atraumatically, but times are shortened by Ca^{2+} from the sea water, and thromboplastic factors from tissue injury. There appears to have been a reduction of clotting time, through the evolutionary series, and fish with higher blood pressures (Chapter 3) have shorter clotting times. Ogilvy and DuBois (1982) comment on the ease with which elasmobranchs bleed from the skin when held out of water with the tail pendent. This does not occur with fast-swimming teleosts such as bluefish, the peripheral vessels of which appear to be stronger. We noted in Chapter 1 that bluefish can generate forces equivalent to 3*G* when they accelerate rapidly. It may be that it is such forces acting on the blood vessels which have led to the strengthening of their walls, a necessary step in the evolution of air breathing and emergence onto land. Whilst much remains to be learnt about haemostatsis in fish the topic is important; blood pressure, the wall strength of blood vessels and blood clotting times are interrelated in complex ways which at present are only partly understood.

Blood viscosity

In Chapter 1 it was emphasized that resistance to flow is due primarily to the viscous properties of the blood, and its viscosity, η, is an important determinant of its ability to act as a transporter of the various agents outlined above. Blood, it was noted, behaves as a non-Newtonian fluid and it is timely to look further at the way its viscosity changes with velocity of flow, i.e. shear rate, with temperature and with haematocrit. The units of η and of shear rate were described in Chapter 1.

The relation of η to temperature

Blood viscosity increases with decline in temperature; the blood of winter flounder, at -1 °C, and subject to a shear rate of 2.3 sec^{-1}, has a η value of over 100 cp, but at 20 °C η is only 27 cp (Graham and Fletcher 1983). At the summer temperatures prevailing in the south of its distribution range, η is approximately 30 cp. The blood of trout has a much lower temperature dependence and at 0 °C and a shear rate of 2.2 sec^{-1}, $\eta = 16$ cp (Fletcher and Haedrich 1987). Hughes *et al.* (1982) studied the passage time of blood of the yellow tail, through Nucleopore filters of known pore size, and found that it increased at lower temperatures; they suggested that the elastic properties of the cells were altered. The greatest change occurred between 10 and 20 °C, and the reduction between 25 and 37 °C was much less. In mammalian red cells two membrane proteins, glycophorin and spectrin, are known to be importantly concerned in their deformability, and whilst it is not known that these occur in fish red cells, it is likely that the great differences between flounder and trout blood cited above reflect, in part, differences in their red cell membranes.

There is some evidence that hypoxia increases the deformability of trout red cells and decreases their passage time through Nucleopore membranes (Hughes and Kikuchi 1984), despite the fact, mentioned earlier, that hypoxia causes the cells to swell. The cause of this remains uncertain. These studies are relevant because in tissues the internal diameter of capillaries is less than that of the red cells and their movement through the capillary lumen is achieved only by their deformation.

The relation of η to haematocrit

Haematocrit varies not only between species but also between individuals; in the 40 trout sampled by Fletcher and Haedrich (1987) it ranged between 24 and 51%. Handling stress in fish increases haematocrit greatly. Elevated haematocrit increases viscosity and in the blood of winter flounder at 0 °C, the transition from a haematocrit of zero to one of 70%, at a shear rate of 2.3 sec^{-1}, leads to an increase of η from 52 to over 200 (Graham and Fletcher 1983). The dependence of viscosity on haematocrit in trout blood is less than that in the winter flounder, but is evident throughout the range of temperatures and shear rates examined.

The relationship of η to shear rate

In the winter flounder the η of both plasma and whole blood shows a marked dependency on shear rate. As this decreases from 90 to 2.3 sec^{-1} the viscosity of whole blood increases from approximately 2 cp to 30 cp at 10 °C, and that of plasma from 2 to 20 cp (Graham and Fletcher 1983). The bloods of Arctic fish such as Arctic char (*Salvelinus alpinus*) and shorthorn sculpin (*Myoxocephalus scorpius*) show this shear dependency much less (Graham and Fletcher 1985, Graham *et al.* 1985), an adaptation, the authors suggest, to the low temperatures to which these species are exposed. Shear dependency is in part due to the tendency of the larger plasma protein molecules, such as those of fibrinogen and the larger globulins, to act as bridges between adjacent red cells causing them to aggregate together to form rouleaux. This is facilitated by low flow rates. Even in plasma alone, molecules of large proteins change their orientation and align with the direction of flow. In the blood of winter flounder plasma accounts for up to 50% of the total viscosity, depending on the concentration of plasma protein in it but this dependence is much less in trout blood (Fletcher and Haedrich 1987).

Graham and Fletcher (1983) have used the morphometric data of blood vessel dimensions in the branchial circulation of the lingcod supplied by Farrell (1980b), to calculate the shear rates in its component vessels. These are presented in the last column of Table 3 (Chapter 3). In narrow vessels such as capillaries and the spaces between pillar cells in the secondary lamellae, shear rates are high for there is only the thinnest of films of plasma between the red cell membrane and the vessel wall. In the gill lamellae shear rates of over 2000 sec^{-1} are calculated to be present and they are high also in the afferent lamellar arteriole. The inverse relation between shear rate and η ensures that viscosity is less limiting in small vessels than in large ones. In the central veins, where velocity and shear rates are low, viscosity will rise and resistance to flow must be greater than their wide diameter suggests. Moreover, lower temperature will further increase η, and will also lower cardiac output. Graham and Fletcher (1983) calculate that with a 10 °C drop in the water temperature of the winter flounder, from 15 to 5 °C, a 50% reduction in cardiac output will combine with the direct effect of temperature on η to cause it to increase at least fivefold. However, such changes are likely to happen gradually, and acclimation may off-set this to some extent by changes in the lipid composition of the red cells.

The leucocytes

Fish have a far higher white cell count than man, and Mulcahy (1970) gives the figure of 798 000 to 137 000 mm^{-3} for the blood of pike. Some seven types of leucocytes are recognized as occurring in fish blood, i.e. three types of eosinophilic granulocyte, a heterophilic/neutrophilic granulocyte, lymphocytes, monocytes and thrombocytes. Plasma cells are not regarded as some kind of white cell as they are not normally found free in fish blood. The lymphocytes are the most abundant white cells and in the spotted dogfish comprise 19%, in the pike 63–93% of the total, again higher than the 15–60% in human blood. They are non-phagocytic: their main role is antibody production. Goldfish acclimated for three weeks at 5, 15, 25 and 35 °C have decreasing numbers of leucocytes at the lower temperatures and their capacity for lymphocyte proliferation is diminished at 5 °C (Dunn *et al.* 1989).

During sexual maturation in trout and salmon, there is a hyperplasia of the interrenal tissue, and an increase in plasma cortisol, which is the most important of the corticosteroids in fish. This brings about a reduction in the number of lymphocytes and the ensuing lymphocytopaenia has been linked, via lowered levels of antibodies, to the increased susceptibility to disease of spawning salmonids (Pickering and Pottinger 1987a). Lympho- cytopaenia is also likely to occur in underyearling salmon parr and is associated with the high, initially density-dependent, mortality during their first three months (Pickering and Pottinger 1988). Such fish are likely to be stressed by starvation, but almost any kind of stress, such as the overcrowding that is likely to occur in aquaculture, leads to lympho-cytopaenia, and death from bacterial and fungal infections (Pickering and Pottinger 1987b).

The granulocytes of the spotted dogfish have been the subject of a number of studies; three types of eosinophil, G1, G3 and G4, are recognized (Hunt and Rowley 1986). G1 granulocytes have an irregular eccentric nucleus and large spherical electron-dense granules; they comprise some 5% of the leucocytes. G3 have a peripheral lobate nucleus and numerous elongate electron-dense granules; they comprise 4.6% of the leucocytes. G4 granulocytes are much the most abundant; they comprise 25% of the leucocytes, have a large nucleus with only a thin rim of cytoplasm and angular electron-dense granules.

G2 are heterophilic/neutrophilic cells with small colourless granules. They have a horse-shoe shaped nucleus, are scarce in stained smears, and comprise only 0.7% of the leucocytes in dogfish blood. In salmonids neutrophils are reported to be the only granulocytes to occur free in the blood and they comprise 1–9% of the white cell count. Doggett and Harris (1989) found neutrophils to be the most abundant of the three types of granulocyte they recognized in the blood of tilapia, and observed them to be phagocytic. In teleosts, but not in elasmobranchs, they are an important part of the mobile defence force and are transported by the blood to the site of injury or pathogen invasion. They can migrate through capillary walls and engulf bacteria and destroy them; their number increases greatly in response to infection or stress. In higher vertebrates neutrophils possess an iodinating system for killing ingested bacteria, which involves peroxidase, hydrogen peroxide and a halide. In dogfish the granulocyte corresponding most closely to a neutrophil contains peroxidase (Parish *et al.* 1986), as do the neutrophils of some primitive teleosts and most of the advanced families including the salmonids (Hine and Wain 1988a). Some primitive groups such as sturgeons have peroxide in their eosinophils.

In mammals serious infections give rise to the presence of toxic neutrophils, characterized by cytoplasmic basophilia and a particular type of granulation. Hine and Wain (1988b) report that typical toxic neutrophils occurred in the blood of the Australian eel when it was stressed by injecting bacterial lipopolysaccharide.

Thrombocytes, we have noted, are important in haemostasis, playing much the same part that platelets do in higher vertebrates. They are spindle-shaped or irregular cells with a large spherical nucleus, and comprise 40% of the leucocytes in the spotted dogfish. In this species thrombocytes are actively phagocytic and can sequester latex beads and carbon (Hunt and Rowley 1986). Monocytes have a round to kidney-shaped nucleus and profiles of endoplasmic reticulum can be recognized in the cytoplasm; they comprise 5% of the leucocytes. Monocytes too are avidly phagocytic and take up injected particles and bacteria.

The generation of new blood cells, i.e. haemopoiesis, and the removal by phagocytosis of old and effete cells are processes which are carried on in a number of tissues and organs: together they maintain an appropriate balance of red cells and leucocytes. Phagocytosis, in addition to removing effete blood cells and tissue debris, is an important non-specific defence mechanism. It is carried out not only by certain sorts of leucocytes – the monocytes and thrombocytes, and in teleosts, the granulocytes as well – but also by the many fixed macrophages which line the vessels of the various haemopoietic tissues. Such phagocytic cells are regarded as constituting a cell system, as they share features of morphology, origin and function in common. It is termed the mononuclear phagocytic system, MPS; it, and the organs of haemopoiesis will form the subject matter of our next Chapter.

5

HAEMOPOIESIS AND PHAGOCYTOSIS –
THE MONONUCLEAR PHAGOCYTIC
SYSTEM

When carbon particles are injected into the circulation of plaice kept at 5–10 °C, 80% of them are removed from the blood within the first 30 min. Labelled red cells of turbot are cleared even more completely; 90% are removed from the blood within the half hour (MacArthur *et al.* 1983). Avtalion (1981) reports that 99% of injected *Staphylococcus aureus* are cleared from the blood of snapper (*Lutianus rivularis*) in 30 min. In the spotted dogfish, injected carbon particles and latex beads disappear from the circulation within 12 h (Hunt and Rowley 1986). Examination shows that localized masses of vascular tissue in the gills, the kidney and the spleen are visibly blackened by the carbon particles. The cells involved in this clearance, and the organs in which they are located, have been the subject of much recent study and deserve our further attention. Some, but not all of these organs are able also to gener- ate blood cells, one or another favouring erythropoiesis, granulopoiesis or lymphocytopoiesis. The two functions, of removing effete blood cells and replacing them with new ones, are so regulated as to maintain a volume of blood sufficient to fulfil its multiple transport and defensive roles set out in Chapter 4.

The phagocytic cells of the MPS system
The monocytes and thrombocytes in the circulating blood can, we have noted, phagocytize particles and bacteria, and may subsequently come to rest on the walls of blood vessels in organs such as the spleen, kidney and cavernous bodies. Thrombocytes appear to be more phagocytic in elas- mobranchs than in teleosts and Ellis *et al.* (1976) report that in plaice they do not accumulate injected carbon particles. The fixed cells of the MPS system that are most involved in phagocytosis are the macrophages.

Macrophages

Macrophages are round cells some 12–20 μm in diameter, with an eccentric round nucleus, a faintly basophilic cytoplasm and numerous granules. Some are stained with blackish and brownish pigments from the iron derived from the breakdown of haemoglobin, a sign that they have phagocytized red blood cells. In the spleen many are seen to contain fragments of red blood cells, indicating the importance of this organ in their removal. In the spleen and kidney, pigmented macrophages, termed melano-macrophages, occur in clusters with which are associated lymphocytes (Herraez and Zapata 1986).

Macrophages can differentiate from monocytes, and those that cluster at sites of injury and inflammation may have differentiated *in situ*. This, at times, makes it difficult to decide how cells intermediate between the two should be named and the term M–Mo cells has been used by Suzuki (1986). Macrophages can be transported in the blood, though they seldom appear in blood smears; they can pass through the walls of blood vessels, especially when they are subject to inflammation. So-called 'fixed macrophages' form interlacing networks which line blood spaces in organs such as the kidney, liver and spleen. The spaces between the musculi pectinati of the atrium (Chapter 2) are, in many teleosts, lined with fixed macrophages; Ellis *et al.* (1976) report that in the plaice, when macrophages have ingested carbon particles, they detach, become free macrophages and probably come to rest in the vascular spaces of the kidney parenchyma.

A separate and important group of macrophages scavenge the walls and mesenteries of the abdominal cavity. In the elasmobranchs the cavity communicates with the outside world through a pair of prominent openings, the abdominal pores. It is less well known, because they are hidden by the cloacal lip, that these occur also in trout, salmon and white fish (*Coreogonus*) (George *et al.* 1982). When carbon particles or bacteria are injected into the abdominal cavity, strings of mucus and macrophages containing carbon particles are later extruded from these openings. We noted in Chapter 2 (Table 1) that perivisceral fluid has a particular ionic composition which differs from that of plasma and sea water. This extrusion of engorged macrophages has not, perhaps, been taken sufficiently into account by fisheries pathologists, for the abdominal cavity is sometimes used as the injection site for immunizing fish against various diseases and the extrusion of the antigen through the abdominal pores may make it difficult to calculate the quantity injected. When irritants such as liquid paraffin are injected into the abdominal cavity, the exudates these provoke come to contain not only macrophages, but monocytes, neutrophils and eosinophils derived from the blood and all of these can be seen to have phagocytized the foreign material (Suzuki 1986). The eosinophilic granulocytes of such exudates in the Australian eel contain peroxidase (Hine and Wain 1989).

Macrophages show a well developed chemotaxis and move through tissues towards sites of injury; Weeks *et al.* (1986) report that those from the blood of fish in highly polluted streams have their capacity for chemotactic migration greatly reduced. Fänge and Pulsford (1985) note that in the thymus of the angler fish many macrophages occur and are seen to be ringed with closely apposed lymphocytes. In the macrophage clusters in the spleen of the spotted dogfish, macrophage cell membranes may be minutely interdigitated with those of the surrounding lymphocytes and in some a complete fusion of the cytoplasm of the two cells occurs (Pulsford *et al.* 1982). The biological significance of this association is uncertain; it may be, as the authors suggest, the occasion when antigens retrieved from phagocytized pathogens are presented to lymphocytes, a necessary stage on their pathway to immunological competence. Antigen-trapping is certainly an important role of macrophages in higher vertebrates.

Haemopoietic and phagocytic organs and tissues
The thymus

In many teleosts the thymus is an opaque, whitish organ which lies dorsally above the gills on each side, just beneath the surface of the opercular lining. Developmentally, it is an epithelial ingrowth from the third gill pouch and the thymus of fish is primitive in that it retains this connection with the surface tissues. Commonly, it involutes once the fish has attained sexual maturity but in some species, such as the angler fish, it persists.

It is bounded with a capsule of collagenous fibres which extends inwards and divides it into lobes. The predominant cells, particularly in young fish, are lymphocytes of various sizes. Plasma cells are also present; they are rounded, have an eccentric nucleus, basophilic cytoplasm and abundant endoplasmic reticulum, and stain intensely with pyronine. In older fish macrophages become more abundant. It is richly vascularized by vessels which enter with the connective tissue partitions and divide to form many capillaries (Fänge and Pulsford 1985). In tilapia, the thymus is partly divided into three zones. A narrow outer zone of immature cells termed thymoblasts surrounds a middle zone of densely packed lymphocytes. The inner zone contains lymphocytes intermixed with macrophages and epithelial cells. Some of these congregate together to form spherical masses termed Hassal's corpuscles. Others enlarge and come to contain vacuoles (Sailendri and Muthukkaruppan 1975). The primary function of the thymus is the production of lymphocytes and it is not concerned with erythropoiesis.

In mammals the thymic or T lymphocytes are quite distinct from the B lymphocytes derived from bone marrow. T cells are responsible for cellular immunity whereas B cells give rise to plasma cells which liberate circulating immunoglobulins. Fish do not have bone marrow but Zapata (1979a) presents evidence that its role in lymphocyte production is taken by the pronephric kidney. There is some evidence that in fish also, two sorts of lymphocytes exist. Mammalian T and B lymphocytes have slightly different densities and can be separated on gravity gradients such as a Percoll gradient. Lymphocytes from the blood of brown trout (*Salmo trutta*), subject to such a separation, form two distinct groups of different density, although the possibility remains that these are developmental stages rather than two groups of cells of different lineage (Blaxhall and Hood 1985). In mice, human gamma globulin (HGG) is known to be an antigen which depends on thymus-derived lymphocytes if antibody to it is to be produced, whereas the production of antibody to the bacterium *Aeromonas salmonicida* is not thymus-dependent but is generated from B lymphocytes. When thymectomized trout are injected with HGG their production of HGG antibody is significantly less than in controls, whereas the production of antibody to *A. salmonicida* is in no way impaired (Tatner *et al.* 1987). This suggests that the equivalent of mammalian T and B lymphocytes exist in fish too; the topic was discussed in detail by Cumber *et al.* in 1982 but the question is still controversial.

The spleen

The spleen is the most important of all the haemopoietic organs. Hagfish have no spleen, but islands of haemopoietic tissue are arranged around the portal vein in *Myxine* and *Eptatretus* (Fänge 1986), and in lampreys, spleen-like tissue is found in the spiral valve of the intestine (Hart *et al.* 1986).

In elasmobranchs the spleen is an elongate, lobed strip of tissue in the mesentery between the stomach and duodenum; it is supplied with arterial blood from branches of the coeliacomesenteric artery and drains into the portal vein. In teleosts the shape and size of the spleen vary greatly and it tends to be more compact than in elasmobranchs. As in higher vertebrates it consists of masses of white and red pulp surrounded by small vessels of arteriolar and capillary dimensions. Structures called ellipsoids occur, which derive from the ends of arterioles and consist of fusiform capillaries sheathed by haemopoietic tissue. The granulopoietic white pulp appears as scattered islands of tissue amongst the red pulp and contains lympho-cytes, monocytes, granulocytes, macrophages and plasma cells. The maturation of granulocytes occurs in the organ or tissue in which they develop, rather than in the blood. The red pulp contains erythropoietic tissue in which the maturing red blood cells can be recognized. Many immature red cells occur in the blood and in fish completion of red cell maturation probably occurs in the vessels (Pulsford *et al.* 1982).

When goldfish are injected with indian ink, carbon particles appear phagocytized into the macrophages of the ellipsoids after only 6 h (Herraez and Zapata 1986). When carp are injected with the bacterium *Aeromonas hydrophila* the antigen first appears in the ellipsoids and from day 15 onwards is localized on the outer surfaces of melano-macrophage clusters of the splenic pulp. Plasma cells, which produce antibody, become more pyroninophilic and abundant and from the 7th–9th day, anti-*A. hydrophila* immunoglobulin appears around them and around the melano-macrophage clusters; it is abundant by the 45th day (Lamers 1986).

The interest that attaches to the spleen as a haemopoietic and phagocy-tic organ should not make us lose sight of its role as a reservoir of plasma and red cells. Strands of innervated smooth muscle encircle and run through it and enable it to contract. The salmon spleen may contain as much as one quarter of the blood volume (Fänge 1984). Contraction of the spleen augments the haematocrit and the oxygen capacity and increases the volume of the circulating blood. The autonomic control of its reservoir function is considered further in Chapter 9.

The gut-associated tissue

In mammals the gut-associated lymphoid tissue (GALT), particularly the Peyer's patches, are an important part of the defence mechanism directed against antigens that enter the alimentary tract. The epithelium overlying the Peyer's patches contains specialized cells, the M cells, which transfer antigens to macrophages and lymphocytes. The gut wall of elasmo-branchs and teleosts includes patches of tissue containing lymphocytes and macrophages (Zapata 1979b), and more recent studies suggest that these may have a similar role. In most, if not all, teleosts the posterior gut is specialized for the uptake and processing of antigens and contains about four times as many intraepithelial macrophages per enterocyte as the first segment. Moreover, their abundance increases when the tissue is challenged locally by antigens. The region has been termed the second gut segment. Rombout and van den Berg (1989) traced, in carp, the passage of both soluble (ferritin) and particulate (*Vibrio anguillarum*) antigens from the gut contents, via epithelial cells to these large intraepithelial macrophages which finally presented antigenic determinants on their outer surfaces as do the M cells of mammals. Other smaller macrophages, loaded with ferritin, moved out of the gut tissue and were later observed in the spleen. The large intraepithelial macrophages may play a part in a local immunity and the small ones may induce a systemic immune response.

The kidney

In the myxinoids, haemopoietic tissue is aggregated around the nephrostomes of the pronephros, and similar clusters of cells occur more posteriorly (Fänge 1984). In the primitive chondrostean, the paddlefish (*Polyodon spathula*) the structure of the kidney does not change suddenly along its length, but the rostral end is largely composed of haemopoietic and MPS tissue and may represent a remnant of the pronephros (Georgi and Beedle 1978). In teleosts the pronephros functions as a kidney only in early larval life; its tubules degenerate as their function is taken over by mesonephric tubules which develop more posteriorly.

The pronephric tissue is replaced by follicles enclosing blood vessels and blood sinuses around which are packed granulocytes, erythrocytes, lymphocytes and plasma cells. Clusters of melano-macrophages occur in the pronephros and are scattered in the renal parenchyma of the mesonephric kidney; they are more intensely melanic than those of the spleen. When challenged with bacterial antigens, lymphocytes appear amongst the macrophage clusters (Herraez and Zapata 1986). Imprints of teleost pronephros show numerous immature red cells, and various sorts of leucocytes, plasma cells and macrophages. In teleosts, such as plaice, the kidney has the greatest uptake of labelled particles of any organ, for it is five times as large as the spleen (MacArthur *et al.* 1983), but comes second to this when uptake is expressed per gram of tissue.

The organ of Leydig, the epigonal organ and the meninges

In elasmobranchs, masses of loose, knotty white tissue occur above and below the oesophagus extending from the end of the pharynx to the cardia of the stomach (Fänge 1968). This structure, the organ of Leydig, lacks erythropoietic tissue but contains many eosinophilic granulocytes, neutrophils, small lymphocytes and plasma cells. It is located in the submucosa between the muscular layer and the mucosal epithelium; a few arteries divide into many capillaries within the lobes and in some species give the tissue a pinkish colour (Fänge 1984).

A mass of similar tissue occurs in some elasmobranchs round the gonads and has been termed the epigonal organ (Fänge 1984). It too has a preponderance of granulocytes and in the spotted dogfish these have been shown to be rich in acid phosphatase, an enzyme normally located within lysosomes. Fänge (1984) notes that there is often a reciprocal relationship between the size of the organ of Leydig and the epigonal organ. Species like the blue shark (*Prionace glauca*) and the nurse shark (*Ginglymostoma cirratum*) have a small or no organ of Leydig and voluminous epigonal tissue, whereas the angel shark (*Squatina squatina*) has the size relationship of the two reversed.

Situated within the cranial cavity of the white sturgeon (*Acipenser transmontanus*), the bichir and some elasmobranchs is a saddle-shaped mass of greyish tissue which is placed above the medulla and the rostral part of the spinal cord (Fänge 1986). It is lobulated and the lobes contain blood and plasma-filled sinuses. Like the organ of Leydig and the epigonal organ, its tissue contains many eosinophilic and heterophilic granulocytes, but it also contains numerous small and medium lymphocytes, and mature and immature erythrocytes. In some specimens the meningeal tissue is pinkish on this account.

The cavernous bodies

In the elasmobranch fish the paired gill filaments of each gill arch are separated by a diaphragm or septum which divides the branchial region into 5–7 gill pouches. The filaments are attached to this diaphragm and in the basal third of it, a column of specialized vascular tissue is interposed between the afferent filamentar arteries and the afferent lamellar arterioles (De Vries and De Jager 1984). In the stringrays (*Urolophus mucosus* and *U. paucimaculatus*) the cavernous bodies of consecutive filaments are partly fused to form a continuous strip (Donald 1989). Each cavernous body contains vascular spaces spanned by trabeculae with a core of connective tissue and smooth muscle, and the walls of its blood spaces, and of the trabeculae, are lined with a basal lamina to which are attached specialized phagocytic cells termed cavernous body cells. They are round and larger than the flattened endothelial cells of the arteriole; they have numerous caveolae, pinocytotic vesicles and many inclusions such as

Table 7. *The relative weights of some haemopoetic organs in fish; % of body weight*

Species	Spleen	Organ of Leydig	Meningeal tissue
Spotted dogfish *Scyliorhinus canicula*	0.58	0.26	—
Blue skate *Raja batis*	0.45	0.18	—
White sturgeon *Acipenser transmontanus*	0.12–0.36	—	0.06–0.07

Data from Fänge, 1968, 1986.

lysosomes, oval electron-lucent vesicles and vacuoles with rod-shaped crystalline structures in them.

These cells phagocytize injected particles such as carbon, latex beads and sheep red blood cells (Hunt and Rowley 1986). Carbon can be seen in the pinocytotic vesicles shortly after its injection and is later transferred into larger vacuoles. Cavernous body cells are not truly members of the mononuclear phagocytic system for they are derived from endothelial cells and are thus mesenchymal in origin.

The cavernous bodies are strategically placed to filter from the blood, prior to its passage into the secondary lamellae, particles and pathogens that might have entered it from the skin or intestine. But they may have other, non-phagocytic, functions such as imparting rigidity to the gill filaments (De Vries and De Jager 1984).

We do not as yet know how the different aspects of haemopoiesis are divided between these various tissues or how the activity of one is related to another. The spleen, it seems, is primarily responsible for erythropoiesis, whereas the organ of Leydig may be mainly concerned in the production of granulocytes, particularly eosinophilic ones. Table 7 shows the relative weights, expressed as percentage of body weight, of the spleen, meningeal tissue and organ of Leydig in three species.

Fish possess less erythropoietic and granulopoietic tissue than mammals on a per cent body weight basis. In mammals the two together amount to 3% of body weight, compared with a total of 0.6–1% in fish. That there is some overall regulation of haemopoietic and phagocytic activity is suggested by the study of Fänge and Johansson-Sjöbeck (1975). Delta-aminolevulinic acid (ALA) is an intermediate in the synthesis of protoporphyrin and haem, and the enzyme δ-aminolevulinic acid dehydratase (ALA-D) is important in the synthesis of haemoglobin. It is thus a useful marker for erythropoietic activity. When the spleen was removed in the spotted dogfish, the haematocrit fell in the first 1–2 weeks, but after three weeks was almost restored as was the ALA-D activity of the blood. The levels of ALA-D in the kidney, epigonal organ and organ of Leydig were all significantly greater than in controls. Histological observation showed enhanced granulopoietic activity in the organ of Leydig and the epigonal organ. These structures, it seems, had taken over in part the erythropoietic function of the spleen.

The blood system serves to link these tissues and organs together. Phagocytosis is an ancient mechanism; amongst the rhizopodous protozoans, and in all of the Phylum Porifera, the sponges, it is the sole means of food intake as well as of defence. All of the external and internal lining cells of a sponge are capable of phagocytosis. Intracellular digestion is important also in the lamellibranch molluscs and the echinoderms. The verbebrates have abandoned phagocytosis as a means of ingesting food and depend on extracellular digestion in an alimentary canal but phagocytosis has persisted as the chief defence against invading organisms and as an essential part of the mechanism of tissue replacement and repair. Many types of cell retain the ability to phagocytize particles on occasion; endothelial cells and the pillar cells of the gills have occasionally been reported to take up injected particles, and the lining cells of the cavernous body, we have noted, are strongly phagocytic. From such simple and generalized beginnings, the system has, it seems, advanced by the evolution of two further mechanisms, both dependent on the blood system. A special type of leucocyte, the monocyte, with the capacity to differentiate into a macrophage, made it possible for greater phagocytic power to be focussed at a site of infection, and for a resident population of macrophages to be concentrated in tissues and organs where they could screen the circulating blood. The macrophages could also collect and process antigens and pass them on to lymphocytes for the formulation of antibodies. Hunt and Rowley (1986) note that in elasmobranchs the induction period for mounting a humoral response is lengthy and, at least in the short term, of questionable value. The nonspecific defence mechanisms of phagocytosis is probably of greater importance in these more primitive fish. In teleosts, when the humoral and cellular immune responses are suppressed by chemicals or low temperature, there is an enhanced phagocytic activity, suggesting that the one compensates for the other. It is in the more recent teleosts that we see evidence of the equivalent of thymus-dependent and thymus-independent lymphocytes.

6

CIRCULATION THROUGH SPECIAL REGIONS

I The microcirculation of the gill

Introduction

The gill of a fish is a multifunction organ. Across its epithelium occurs an exchange of the respiratory gases and in most fish it is the major if not the sole organ of gas exchange. It is also a major site of water ingress and egress, by osmosis. It is significantly involved in ionic regulation, and in acid–base balance. It is an important site of nitrogen excretion. Toxicants ingested with the food may be cleared through the gills; Nakatani (1966) reports that most of the zinc ingested by trout is lost through the gills and Matthiessen and Brafield (1977) suggest that this occurs through the activity of the chloride cells. There is now evidence that the microcirculation of the gill is specialized in connection with some of these functions. Although there are differences between species, three separate circulatory pathways can be recognized in the fish gill.

On each side of the pharynx of a teleost are four gill arches, each bearing a double row of gill filaments; the tips of the filaments of adjacent arches touch, so that water has to flow between adjacent filaments. Each bears, on its upper and lower surfaces, a row of closely spaced leaflets, the secondary lamellae. A secondary lamella can be likened to an envelope. Its two parallel faces are spaced apart and anchored by pillar cells which prevent it from ballooning out in response to the pressure of the blood within. The lamellar wall consists of two layers; the outer epithelial layer may be one or two cells thick and is supported by a strong basement membrane. The inner layer consists of the flange of the pillar cells, for these are shaped like spools and their widely spread flanges have their edges joined by desmosomes.

The arterioarterial pathway

The arterioarterial pathway (Laurent and Dunel 1976), carries blood concerned in gas exchange and in most fish receives the entire cardiac output which, from the ventral aorta, passes up the four or five pairs of afferent branchial arteries alongside their respective branchial arches. From these an afferent filamentar artery passes down each gill filament (Figure 19), and gives rise to muscular afferent lamellar arterioles which supply blood to the secondary lamella; in the distal third of the filament some of these arterioles supply two lamellae. We have already described, in Chapter 5, the structure of the cavernous bodies interposed between the afferent filamentar arteries and the afferent lamellar arterioles in elasmobranch gills. Their importance as a phagocytic tissue intercalated into the arterioarterial pathway demanded their earlier consideration.

Blood flows between the walls of the lamellae by percolating between the pillar cells and along a marginal channel free of pillar cells, which encircles its outer edge. A similar channel occurs along the base of the

Figure 19. The main vessels of a teleost gill. A.F.A. afferent filamentar artery, A.L.A. afferent lamellae arteriole, E.F.A. efferent lamellar artery, E.L.A. efferent lamellar arteriole, F.V. filamentar veins, Il.V. interlamellar vessel, M.C. marginal channel, Nut.V. nutrient vessel, P.C. pillar cell, S.F.A. subsidiary filamentar artery, S.L. secondary lamella. Drawn partly after Boland and Olson 1979.

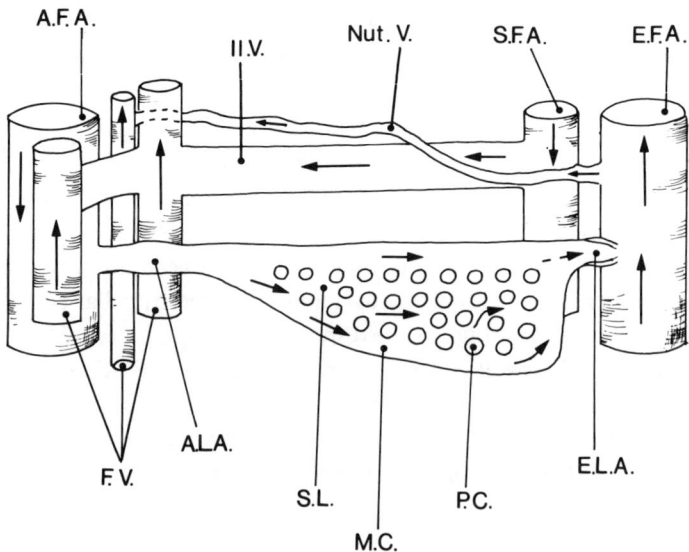

lamella, embedded in the filament. The efferent lamellar arteriole is less constricted and joins to an efferent filamentar artery.

The two cell layers of the lamella wall constitute the water–blood barrier; in the sole (*Solea solea*) this barrier is 1.97–3.98 μm thick, comprised as follows: epithelial layer 1.3–3.4 μm, basement membrane 0.03–0.41 μm, pillar cell flange 0.03–0.18 μm. In the trout gill the total thickness is greater, 6.4 μm. In fast-swimming pelagic fish such as the skipjack tuna (*Katsuwonus pelamis*) it is very much less, 0.598 μm (Hughes and Morgan 1973). Almost all of the reduction in thickness of the lamellar wall in such fish is due to the very thin epithelial layer.

Blood pressure and blood flow through the secondary lamellae

Recent studies have focussed attention on how the blood flows through the secondary lamellae, and their response to an increase in blood pressure. The pillar cells take up some 10% of the volume between the lamellar walls, and the equations of Poiseuillean flow are difficult to apply to the spaces between the pillar cells, which are wider than long. Farrell *et al.* (1980) have treated the problem as one of sheet flow, such as is now known to occur in the alveolar circulation of mammals.

The length and breadth of the lamellae, and thus of the sheet of blood within them, increase very little as intralamellar blood pressure rises, but the lamellae do increase in thickness. This is not because the pillar cell columns stretch (Booth 1979a), but rather that their flanges bulge outwards. In the lingcod the increase in thickness of the sheet is linear with pressure between pressures of 22–44 mm Hg, which approximate to those obtaining in life. The pillar cells contain myosin filaments which extend the length of the pillar cell column. Smith and Chamley-Campbell (1981), on the basis of immunofluorescence histochemistry, conclude that the pillar cells of the golden bream (*Chrysophys auratus*) contains as much smooth muscle myosin as mammalian vascular smooth muscle. It is not innervated and it probably does not contract and maintain a steady tension against transmural pressure but it may, the authors suggest, contract in response to rapid increases in pressure and provide a measure of autoregulation of flow. Bartels and Decker (1985) report that in the hagfish *Myxine* the flanges of pillar cells are joined by gap junctions. These may imply, the authors suggest, that some electrotonic coupling occurs across the surface of the pillar cell layer which assists in the coordinated synthesis and secretion of mucus.

A consequence of the increase in thickness of the sheet of blood with rise in blood pressure is to bring about a redistribution of blood flow through the lamella. At rest most flow occurs in the marginal and basal channels but with rising transmural pressure, flow in the central part of the lamella increases, as does the total area of lamellar surface which is perfused with blood. Farrell *et al.* (1979), in other studies on isolated lingcod gills, report that at rest only the lamellae in the basal two thirds of each gill filament are perfused, and that increased perfusion pressure recruits lamellae that lie towards the distal ends of filaments. This may perhaps be because their afferent lamellar arterioles have higher critical closing pressures. Because, once they have opened, such arterioles remain patent below their critical closing pressure, increased pulse pressure also causes recruitment. This happens when, for example, the heart pumps at a lower rate, with a larger stroke volume, so that total flow is maintained; this occurs during the bradycardia of hypoxia. The recruitment of additional lamellae reduces the total resistance of the gill vasculature. In trout, hypoxia causes a peripheral vasoconstriction and cardiac output is maintained; as a consequence dorsal aortic pressure rises. Soivio and Tuurala (1981) have shown that this causes an increase in the thickness of the sheet of blood within the lamella; it is likely that here, too, this brings about an increased flow through the central parts of individual lamellae, and a recruitment of additional lamellae. This type of response may account for the finding, by Farrell and Daxboeck (1981), that when lingcod are subjected to progressive hypoxia they maintain and even increase their oxygen uptake down to a P_{O_2} of 45 mm Hg. We will have reason to consider these changes again when we discuss the response to exercise (Chapter 10) and to hypoxia (Chapter 11).

Records of transmural pressure in different positions along the arterioarterial pathway (Figure 20), have been made by Farrell and Smith (1981), using micropipettes. *In vivo* studies of lingcod gills show that the input pressure to the afferent arterioles differs rather little along the length of the afferent filamentar arteries for these are vessels of relatively large diameter, and 32% of the total resistance lies between the entrance to the lamella and the ventral aorta. The main resistance is located in the afferent lamellar arteriole and in the lamella itself; these together account for 59% of branchial vascular resistance. The efferent arteriole accounts for only 9% of the total and the authors conclude that the transmural pressure in the lamella is probably not much below that in the dorsal aorta.

The arteriovenous pathway

The tissues of the gill are provided with a nutrient circulation (Figure 19, Nut.V.), analogous to the coronary vessels of the heart and, like them, arising from the efferent gill vessels (Boland and Olson 1979, Olson and Kent 1980). The vessels are very thin, follow tortuous pathways and, it is supposed, carry oxygen and soluble nutrients to the tissues of the gills. They finally empty into filamentar veins which open into the branchial vein in each gill arch. This lies on the outer side of the afferent branchial artery. We should note that the nutrient vessels are not alone responsible for carring oxygen to the gill tissues for, in the cod, some 42% of it diffuses directly in from the water (Johansen and Pettersson 1981), and this statistic doubtless differs from one species to another.

In many teleost fish the chloride cells, responsible, it is believed, for the active transport of ions, are located in rows in the troughs between adjacent secondary lamellae. In other species they may extend onto the lamellar surface. They vary in abundance between 1% in freshwater adapted tilapia to 13% in seawater adapted toad fish (*Opsanus beta*) (Perry and Walsh 1989). These authors present evidence which suggests that chloride cells are metabolically more active than epithelial cells when

Figure 20. Microvascular pressures in the gill vessels of a lingcod recorded *in vivo*. Ordinate, pressure as % of ventral aortic pressure. Abscissa, pressures in afferent vessels, lamellar complex, (L.C.), and efferent vessels. P.V.A. pressure in ventral aorta, P.D.A. pressure in dorsal aorta. The lamellar complex includes the secondary lamella, plus afferent and efferent arterioles. Redrawn from Farrell and Smith 1981.

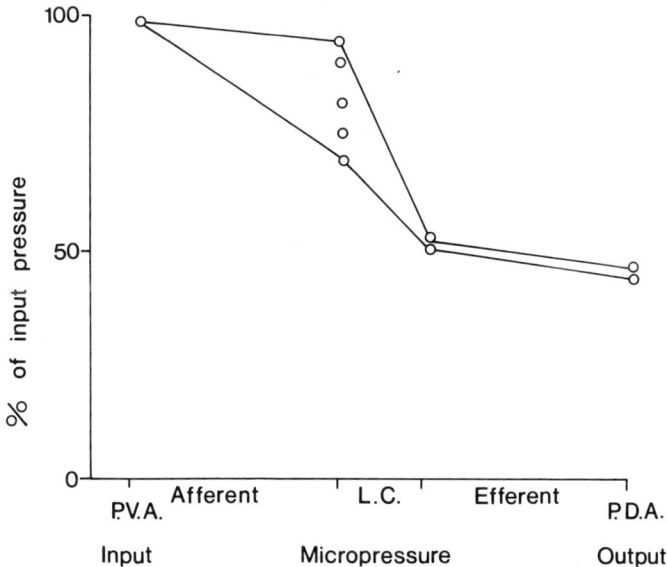

the two are studied in cell suspensions. Chloride cells are more abundant in trout that inhabit waters of low ionic content (Thomas *et al.* 1988). Such fish are less able to resist severe hypoxia because, the authors suggest, a smaller proportion of the gill area is available for gas exchange and chloride cells make heavier demands on the available oxygen supply.

In the channel catfish and spiny dogfish interlamellar vessels underlie these chloride cells and may provide a third circulation specifically concerned with ionic regulation. They are of broader diameter than the nutrient vessels and rise from subsidiary filamentar arteries. These derive from efferent vessels containing oxygenated blood. The interlamellar vessels empty into filamentar veins and thus into the central veins of the head. They are, as we shall see in the next section, properly regarded as vessels of the secondary circulation. The nutrient and interlamellar veins together are termed the arteriovenous pathway. Olson (1984) has examined the outflow from the venous drainage and the efferent artery, of isolated, perfused trout and ictalurid gills. The venous outflow increased with increase of efferent artery pressure; moreover, efferent artery blood always had a higher haematocrit than ventral aortic blood, suggesting that plasma skimming occurs at the openings into the arteriovenous pathway. The corollary to this should be that branchial vein blood has a lower haematocrit and haemoglobin content than dorsal aortic blood. In a careful *in vivo* study of cannulated trout, Ishimatsu *et al.* (1988) have shown that dorsal aortic blood had a haematocrit of 24.3% as against 3.5% in the branchial vein; the corresponding figures for haemoglobin content were 6.5 and 1.0 g 100 ml^{-1}. In such preparations the proportion of the cardiac output directed through the arteriovenous pathway is very much smaller, 7%, than the values derived from isolated perfused gills. One reason for this discrepancy may be that isolated preparations are sometimes pretreated with catecholamines which dilate the openings into the arteriovenous pathway and increase flow through it.

Duff *et al.* (1987) report that, following an injection of labelled cells into the ventral aorta of trout, the red cell space of the gills reaches an early initial plateau after 12 min and then achieves a second later rise at 150 min. These authors suggest that the first corresponds to the perfusion of the arterioarterial pathway, and that the second peak occurs as labelled cells infiltrate the small vessels of the nutritive and interlamellar circulations.

The arrangement of vessels outlined above applied to some but not all gills and there are many differences in published accounts. The numereous small branching vessels which underlie the lamellae constitute a vascular bed of great complexity. Olson (1983) has shown that in the trout, the appearance of this vasculature in scanning EM photographs depends very much on the pressure with which the polymer mass is injected. At low pressures the vascular bed has the appearance of many parallel vessels; as pressure increases it approximates more and more to a continuous sinus. In eels, and in stingrays, there really is a central venous sinus at this site, and both the afferent and efferent filamentar arteries have connections with it (Donald 1989). Its reported presence in some other species may, however, be an artefact of preparation.

There are other variations in the microvasculature of this region in different fish. In the black bullhead (*Ictalurus melas*) and the channel catfish, there are numerous arteriovenous anastomoses (AVAs), which run from the afferent arterioles to join the middle of the interlamellar

Figure 21. The secondary blood system of teleost fish. *A*, Diagrammatic figure of part of the secondary system of the glass catfish. *B*, Origin of the interarterial anastomoses in the gourami *Osphronemus goramy*. *C*, Opening into an anastomotic vessel in the trout, showing the microvilli. C.V. caudal vein, D.A. dorsal aorta, I.A.A. interarterial anastomoses, Sc.A. secondary artery, Sc.C. secondary capillaries of anal fin, Sg.A. segmental artery, Sg.V. segmental vein, Sc.V.F. secondary vein at base of fin, Sv.V. subvertebral vein, Vt. vertebra. Redrawn, *A* from Steffensen *et al.* 1986, *B* from Vogel 1981, *C* from Vogel 1978.

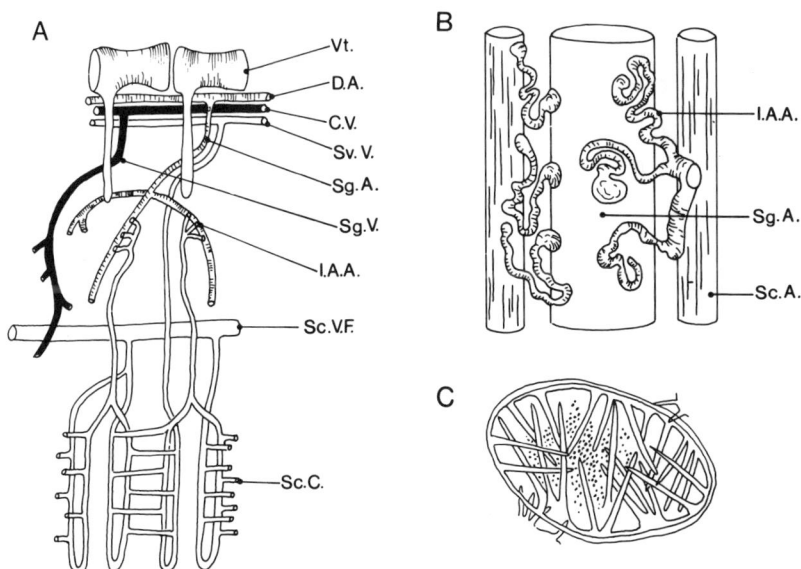

vessels; such AVAs occur also in the dogfish and in the bowfin (*Amia calva*) (Olson 1981), but are rare in the trout (Olson 1984).

In the literature, the reader will sometimes find these interlamellar vessels referred to as venolymphatics. Lymphatics in higher vertebrates originate blindly in the tissues. The interlamellar vessels are always connected to arteries and are to be compared with the capillaries of the secondary blood system, the subject of the next section.

II The secondary blood system

Over the last decade, the studies of Vogel and his collaborators (Vogel 1981, 1985a, b, Vogel and Claviez 1981) have shown that there exists in fish a second system of arteries, capillaries and veins which, in part, runs parallel with the primary system so far described. Short interarterial anastomotic vessels arise (Figure 21*A*, *B*) from the efferent branchial arteries, dorsal aorta, and segmental arteries; they are of narrow diameter and are much coiled and twisted. They join to form paired vessels either side of the aorta and segmental arteries (Figure 21*B*) and tributaries of these subdivide to form a network of secondary microvessels, which drain into secondary veins.

Secondary microvessels are not distributed evenly to all tissues; they are

Figure 22. The distribution of the secondary microcirculation to the different tissues of the body. Redrawn partly after Vogel 1985a.

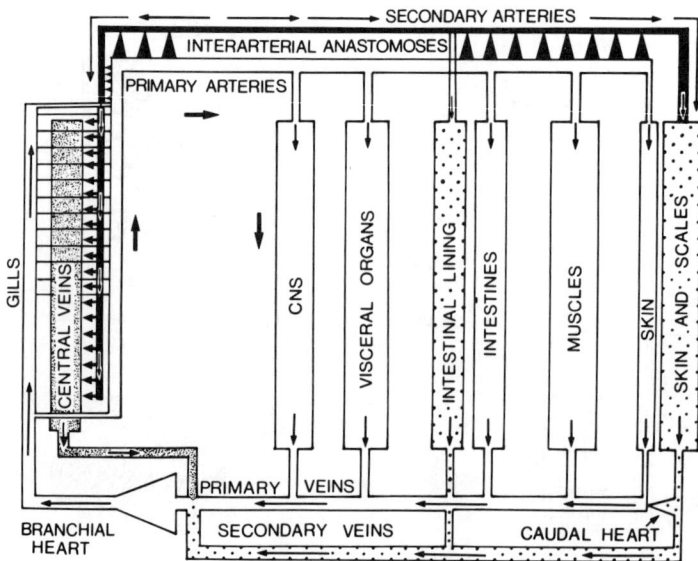

absent from the skeletal muscles, the central nervous system, and the solid organs of the digestive system such as the liver and pancreas (Figure 22). They are abundantly present in the skin where they form a net of capillaries beneath the epidermis of the exposed part of each scale (Figure 23). They occur in the buccal cavity, and to a lesser extent beneath the gut epithelium. The interlamellar vessels of the gills, described in the previous section, are also probably secondary in origin. They differ in that they open into central veins of the head such as the branchial and inferior jugular vessels. Secondary capillaries are virtually confined to epithelial surfaces, and for the most part these face towards the water or into the gut and have the potential to take part in osmotic and ionic exchanges.

The blood from the secondary microvessels of the skin is collected up

Figure 23. The microcirculation of a scale of the pike. Art. afferent arterioles running caudally towards the scale margin, C. capillaries in the exposed part of the scale, Vn. veins running rostrally beneath the overlying scale. Arrows show direction of blood flow. Redrawn partly after Tysekiewicz 1969.

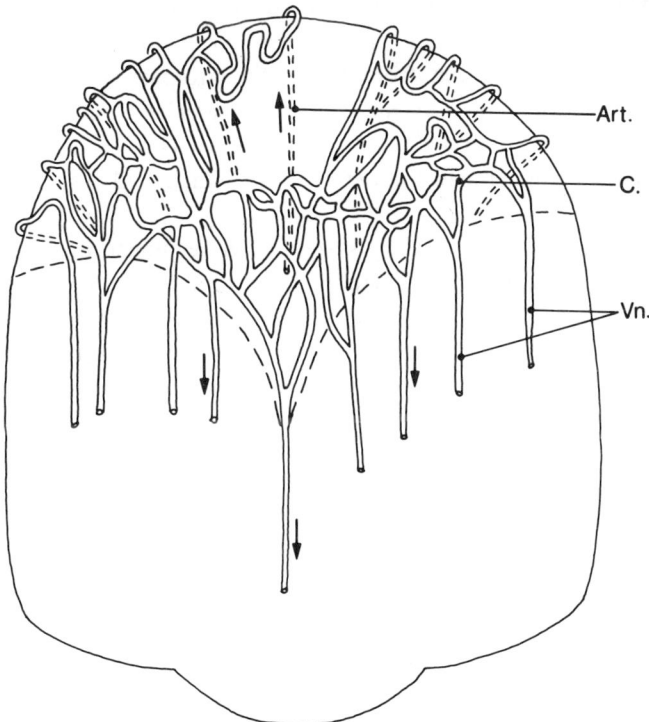

into the dorsal, lateral and ventral cutaneous veins; in addition, one or two prominent secondary veins, the subvertebrals, runs in the haemal canal beneath the caudal vein (Figure 35, below). The cutaneous veins are the chief veins of the secondary system. Blood from them is returned to the primary venous system by an auxiliary venous pump, the caudal heart, a two chambered pump located in the tail fin and powered by skeletal muscles driven from motor neurones in the tip of the spinal cord (Vogel 1985b). Its structure will be considered in Chapter 8.

The openings of the anastomotic arteries, at the sites where these arise from the primary arteries, are, in trout and in tilapia, guarded by a fringe of microvilli (Figure 21C). These appear to be contractile and, in their extended state, they can form a filter which holds back some or most of the red cells (Vogel *et al.* 1974). They are not, it appears, present around these openings in the black bullhead (Olson 1984). When they elongate, the blood in the secondary veins then appears to be clear and without cells. It was the clear appearance of the fluid in these vessels which led earlier anatomists to regard them as lymphatics and the caudal heart as a lymph heart. They cannot strictly be lymphatics, for, as we have noted, they arise from and remain connected with the primary arteries (Vogel and Claviez 1981). Steffansen *et al.* (1986) have observed these anastomoses in live specimens of the glass catfish (*Kryptopterus bicirrhis*), and report that cells can be seen, from time to time, to pass from the primary arteries into the secondary system. Ishimatsu *et al.* (1988), in their study of the haematocrit of the blood in the branchial veins of trout, mentioned above, noted a higher haemoatocrit in fish which had just undergone surgery than in rested fish.

The function of the secondary system is as yet uncertain. Steffansen (unpublished, but quoted by Farrell in correspondence) has determined the plasma volume of the secondary system of trout as 4.8% of the body mass, compared with 2.3% in the primary system. More than twice as much of the plasma, at any one time, it appears, is in the secondary vessels as is in the primary circulation. The calculated flow rate for a 378 g trout is 2.5 ml h^{-1}. Davie (1982) estimated that the caudal heart of an 460 g eel, pumped 0.5–1.0 ml h^{-1}, a figure in good agreement, for the flow through the secondary circulation of the gills would be excluded from this total.

The secondary system may be concerned with ion and water balance. The secondary capillaries of the scales all run in a rostral direction; they emerge from beneath the margin of a scale, loop over it beneath the thin epidermal layer and pass rostrally until they disappear beneath the edge of the scale above (Figure 23) (Tysekiewicz 1969). The skin obtains much of its oxygen, and in some species all of it, by direct diffusion from the water and in such circumstances there is little need for the capillary vessels to contain red cells. Osmotic and ionic exchanges, and the carriage of nutrients, could well be mediated by plasma alone, and this may account for the provision of filtering microvilli across the openings of the anastomotic arteries. Moreover, the exclusion of most of the red cells will lower the viscosity of the blood in the secondary system and these narrow, coiled vessels may impose a resistance and thus loss of pressure which makes this advantageous.

III   The renal portal circulation

In Chapter 3 it was emphasized that in most fish the caudal vein and its continuation in the abdomen, the paired renal portal veins, form the sole highway for the return of blood from the post abdominal axial muscles. Valved segmental veins join it as it passes forwards in the haemal canal (Figure 24), and more rostrally, similar vessels enter the renal portal veins. The blood collected into the renal portal veins thus comprises the drainage from the capillary circulation of the scales, the axial muscles and the nerve cord and is distributed to the capillary networks that invest the renal tubules.

In mammals there is seldom any connection between somatic veins from skeletal muscle, and the hepatic portal system draining the viscera. This ends in the sinusoids of the liver, whereas somatic venous blood passes directly to the heart. Teleosts, in contrast, do have such connections, a feature that may be related to the presence of a remarkable neuro-haemal organ, the urophysis, in the terminal portion of the spinal cord.

Figure 24. The renal portal circulation. The urophysis U, at the tip of the spinal cord secretes the urotensin peptides into the caudal vein. C.V. caudal vein, R.P.C.V. right posterior cardinal vein, R.R.P.V. right renal portal vein, S.C. spinal cord, S.I.V. segmental intercostal veins.

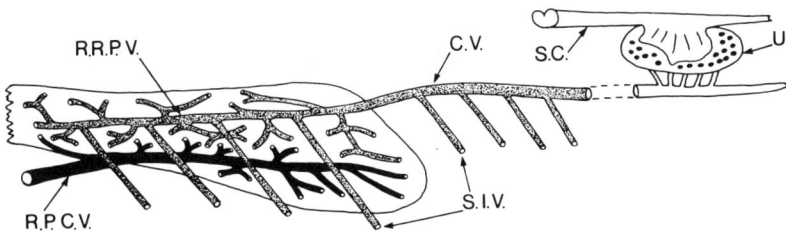

The urophysis

The urophysis contains the terminals of neurosecretory cells which lie in the tip of the spinal cord, and secrete peptide hormones, the urotensins. From it, capillaries transport these into the caudal vein (Figure 24 U.). Urotensins 1 and 2 have been sequenced and synthesized; urotensin 1 is a 41-residue peptide with a striking sequence homology with ovine cortico-tropin-releasing factor. It is known to inhibit salt and water uptake from the isolated post-intestine of the 5% seawater adapted, long-jawed goby (*Gillichthys mirabilis*) (Loretz *et al.* 1983), and seawater adapted tilapia (Mainoya and Bern 1982). Urotensin 2 is a cyclic dodecapeptide with homologies to somatostatin; it stimulates the uptake of Na^+ and Cl^- from the bladder of the seawater adapated goby, from the isolated post-intestine of this goby acclimated to 5% seawater (Loretz and Bern 1980, Loretz *et al.* 1983) and from the intestine of seawater adapted tilapia (Mainoya and Bern 1982). There are other actions of these and other urotensins, and the reader is referred to the review by Lederis (1984) for an account of them. In the molly (*Poecilia latipinna*) other neurones in the urophysis secrete serotonin; they are part of an additional, as yet uninvestigated system, the secretions of which must also pass into the caudal vein (Cohen and Kriebel 1989).

The body fluids of marine teleosts have an osmolarity approximately one third that of sea water. The additional water they require for this dilution is gained by drinking sea water and excreting the salt in it by the chloride cells of the gills. There are ion transporting cells in the intestinal and bladder epithelia. There may thus be advantages in having a direct venous connection which enables the urophyseal peptides to gain access to these epithelia without having to pass through the whole of the circulatory system.

The bladder vein

In the lingcod (Allen 1905) the caudal vein gives rise to a branch to the bladder just prior to its division into the renal portal veins (Figure 25, Bl.V.). It subdivides into a network of small vessels on the posterior–dorsal surface of the bladder, an area which in the long-jawed goby is lined with a special cuboidal epithelium concerned with the active reab-sorption of Na^+ (Loretz and Bern 1980). Indian ink injections into the vein show that its branches ramify amongst the cuboidal cells. On the anterior–ventral surface of the bladder is a second network which joins to a vessel that carries blood, via the gonadial veins, to the right posterior cardinal vein. Connections to the bladder have been described in the sea perch (*Sebastodes*), in the scorpion fish (*Scorpaenichthys*) and the green-fish (*Hexagrammus*) (Allen 1905), and in all three the blood is returned to the posterior cardinal vein.

Connections with the posterior intestine and rectum

In the lingcod, a branch vein leaves the right renal portal vein just after its entry into the kidney (Figure 25, B.P.I.V.), passes through this without branching, and runs ventrally to join the posterior intestinal vein (Allen 1905). A similar branch occurs in the eel, tench and carp. These connections reveal that both the bladder and the intestine have their own portal circulations in parallel with that of the kidney. Whilst these have not yet been subjected to experimental study, the possibility that they provide local circulations for the urophyseal hormones remains an interesting speculation.

Figure 25. Branches of the caudal vein in the lingcod. Bl. bladder, Bl.V. bladder vein, B.P.I.V. branch to join the posterior intestinal vein, C.V. caudal vein, G. gonad, G.D. gonoduct, Int. intestine, K. kidney, P.Int.V. posterior intestinal vein, R. rectum, R.P.C.V. right posterior cardinal vein, R.R.P.V. right renal portal vein, R.V. rectal vein, Ur. ureter. The arteries and segmental intercostal veins have been omitted. Redrawn after Allen 1905.

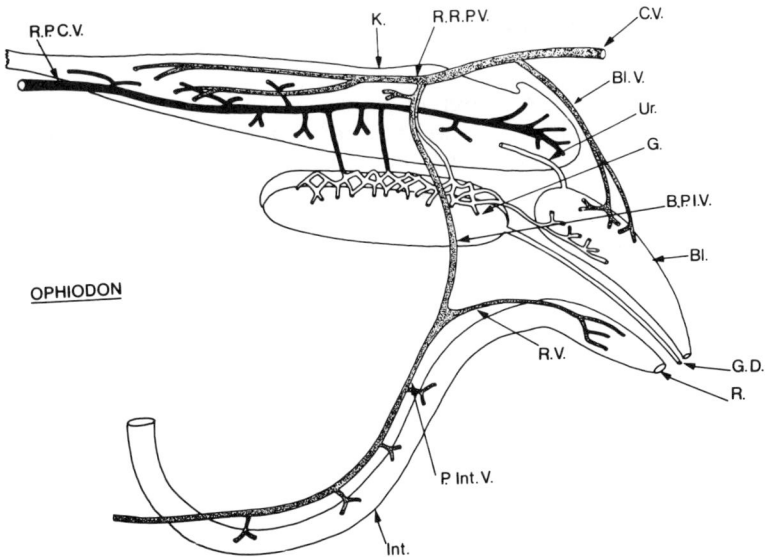

Afferent and efferent flows to the kidney

The teleost kidney is not uniformly structured along its length. There is often a division between the paired anterior lobes, which are rich in tubules but have few or no glomeruli and the median lobe caudal to these, which has many glomeruli clustered together. Such kidneys occur in freshwater species such as eel, tench, stickleback (*Gastrosteus*) and rudd (*Scardinus*) as well as in some marine genera such as the cusk eel (*Ophidium barbatum*) and the red band fish (*Cepola rubescens*).

The renal portal veins run only to the lobed portions of the kidney where they subdivide to form a capillary network around the many tubules. The caudal portion is perfused entirely from the dorsal aorta which branches rather irregularly into many small renal arteries. The glomeruli are very numerous and are arranged in clusters like bunches of grapes. Audigé (1910) termed them '*glomeruli en grappe*'. They receive blood from the numerous small renal arteries which continue on to supply the capillary nets around the tubules. The exclusion of the glomerulus-rich posterior lobe from the renal portal circulation, must imply that it does not receive the urotensin hormones directly, as do the tubules rostral to it.

This subdivision into tubule-rich and glomerulus-rich sections is, we have noted, common in freshwater species, and may reflect the great need to eliminate water taken in osmotically through the gills. This enlargement of the posterior lobe reaches its extreme in the common eel (Figure 28), where it extends backwards from the posterior wall of the abdomen, and lies in a finger-like pocket below the spine. The condition is seen also

Figure 26. The blood vessels of the middle and posterior kidney of the eel. Lettering as in Figure 25; in addition, B.R.P.V. branches of right renal portal vein, D.A. dorsal aorta, G.E.G. glomeruli 'en grappe', M.K. middle kidney, P.K. posterior kidney, R.A. renal artery to posterior kidney, V.T.M. vein from trunk muscles. Redrawn partly after Audigé 1910.

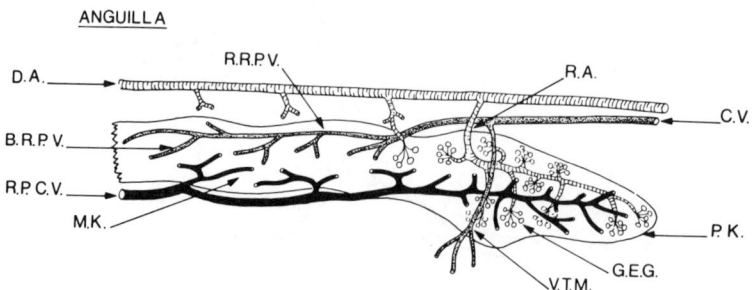

in the pond loach (*Misgurnus fossilis*), and the brill (*Rhombus rhombus*) and turbot (*R. maximus*); these two species can occur in the freshwater parts of estuaries (Audigé 1910). In those species, mainly freshwater or euryhaline, which have a glomerulus-rich posterior lobe, the distinction between this and the anterior tubule-rich portion is further emphasized by the ureters. The primary ureters run posteriorly down the outer side of each lobe of the anterior portion receiving collecting ducts in a regular sequence. The posterior lobe has its own paired ureters which arise from separate outgrowths of the bases of the primary ducts. The rostral portion, rich in tubules, reaches its extreme of development in marine genera such as the angler fish, the sea horse (*Hippocampus*) and the pipe fish (*Syngnathus*). In these the kidney entirely lacks glomeruli and consists only of a single median mass of tubules. It lacks any arterial supply and is perfused only with low pressure blood from the caudal vein.

IV   Red and white muscle

The axial muscles of fish comprise 40–67% of their body weight and power the swimming necessary for food capture, escaping enemies and performing other vital functions. These muscles exist in two main forms, red and white, and they have very different patterns of microcirculation.

In most fish red muscle occurs as a superficial layer on each side (Figure 27*A*), and is concentrated into a wedge beneath the lateral line. In trout it constitutes some 1% of the body mass (Barron *et al.* 1987). It is responsible for sustained swimming, and is red because it has a rich blood supply and its fibres contain myoglobin, which acts as an internal oxygen transport system. Trout subjected to forced swimming for 200 days increase their red muscle mass 2.2-fold (Davie *et al.* 1986). Red muscle metabolizes both lipid and carbohydrate, the latter by the tricarboxylic acid cycle. In fish which swim with their fins, such as the chimaerae, and the skates and rays, the distribution of red muscle is different, but the significant point is that it is concentrated where the muscles for sustained swimming are located. White muscle comprises some 66% of the body mass of a trout and is white because it is sparsely circulated and lacks myoglobin. It provides the main propulsive power for fast swimming of brief duration (Chapter 10) and metabolizes carbohydrate largely glycolytically. The white muscle of the ocean pout lacks the low molecular weight fatty-acid binding protein that is present in its myocardium (Stewart and Driedzic 1988).

Red muscle fibres have many capillaries running longitudinally between them (Figure 27*B*) which give off cross links transversely to the fibres; together these form a capillary network which invests the surface of the fibre (Mosse 1978). White muscle has only a few capillaries (Figure 27*C*), and these run longitudinally; the fibres are not invested with a capillary network. In the sturgeon (*Acipenser stellatus*) (Kryvi *et al.* 1980), and in some other genera, a third type of muscle, pink in colour, lies between the red and white muscle and has properties intermediate between the two. In a mid cross section 76% of the area is occupied by white fibres; pink fibres occupy 5% and red fibres 19%. The number of capillaries surrounding each fibre is: red 2.3, intermediate 0.9, and white 0.2. Sixteen per cent of the surface area of red muscle fibres is covered with capillaries, compared with 5% for intermediate and 1% for white fibres. Cameron (1975) has shown, by injecting labelled microspheres into arctic grayling (*Thymallus arcticus*), that red muscle has 10 times the blood supply of white, calculated on a per gram basis. Nevertheless, white

Figure 27. *A*, Distribution of red and white muscle in the post abdominal trunk of the glass whiting *Haletta semifasciata*. *B, C*, The abundance of capillaries, (*C*) in *B*, the superficial red muscle and *C*, the deep white muscle. Scale bars = 100 μm. Redrawn after Mosse 1978.

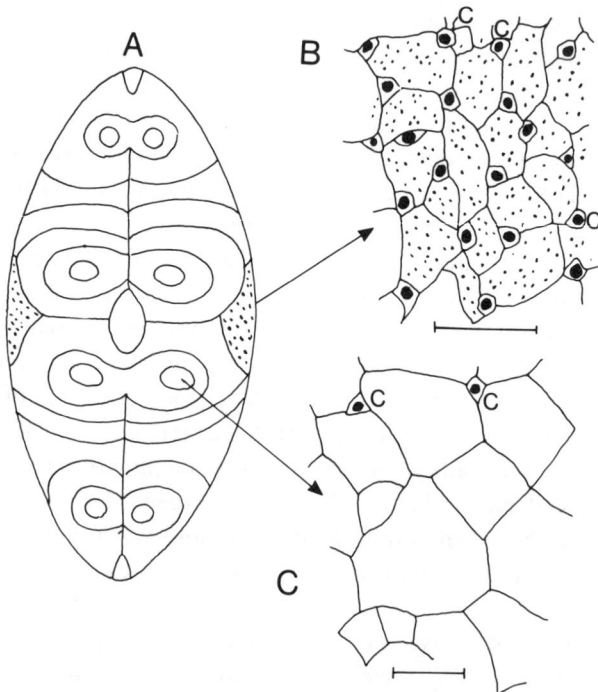

muscle, because of its abundance, receives the largest proportion of the cardiac output of a resting fish; Barron *et al.* (1987) report that in trout at 6 °C the flow to red muscle is 10% of the cardiac output compared with 40% to white, and that with rising temperature the proportion flowing to white muscle increases, to 75% at 18 °C. This increased flow with rising temperature, the authors suggest, reflects a decrease in vessel resistance which prevents the rise in blood pressure that would inevitably follow the rise in cardiac output, from 19.8 ml min^{-1} kg^{-1} at 6 °C to 62.5 ml min^{-1} kg^{-1} at 18 °C.

Red muscle fibres contain many mitochondria and white muscle very few. Expressed in the form of mitochondrial volume as a percentage of fibre volume, the figures for sturgeon are: red 30%, pink 3.7%, and white 0.7%. Closely similar data have been published for the red, pink and white muscle of the velvet belly shark (*Etmopterus spinax*), and of the blackmouthed dogfish (*Galleus melanostomus*) (Totland *et al.* 1981). The richness of the vacularization is directly related to the volume fraction of the mitochondria. It is also related to the rapidity with which metabolites are removed. In crucian carp (*Carassius carassius*), Johnston and Goldspink (1973) have shown that, following forced exercise, the lactate and glucose levels of the red muscle rapidly return to normal, whereas the lactate level in white muscle remains elevated even after 5 h.

The paucity of circulation in white muscle may be compensated by its intimate anatomical relationship with the adjacent red muscle. Wittenberger *et al.* (1975) suggest that there is a non-circulatory transfer of lactate from white muscle to red through a direct coupling of cells. There may also be a reverse transfer of glucose. Whilst this remains uncertain, it is clear that the proportion of red muscle in a fish is an indication of its ability to sustain swimming and that the aerobic capacity of this muscle is enhanced by its rich microcirculation.

We have noted earlier that all blood returned to the heart must pass through the gills. As thermal diffusion is 10 times as rapid as molecular diffusion, it follows that a gill area sufficient to oxygenate the blood will certainly cause it to attain the same temperature as the water which surrounds it. Thus most fish cannot retain the heat generated by their own metabolism and are uniformly poikilothermic. The circulatory rearrangements which enable certain elasmobranchs and teleosts to escape from this limitation form part of our next Chapter.

7

RETIAL COUNTER-CURRENT SYSTEMS: FLOW-DIFFUSION-CONCENTRATION

Introduction

In fish, arteries and veins often run side by side in the restricted spaces around or between skeletal elements or muscle masses. The afferent and efferent branchial arteries alongside the gill arches, the dorsal aorta and posterior cardinal veins beneath the spine, and the caudal artery and vein within the haemal canal are examples of this. Smaller vessels derived from these become similarly juxtaposed and their thin walls may be enclosed in a common connective tissue sheath. Such close proximity may make it possible for agents such as heat and oxygen to diffuse across the walls from one stream of blood into the other.

A characteristic feature of some fish blood systems is the occurrence of masses or slabs of highly vascular tissue, consisting of parallel arrays of arterial and venous vessels, of capillary or arteriolar dimensions. These vessels may have a mosaic or hexagonal packing so that each arterial vessel is bordered by a venous one, and their walls have the maximum possible contact. Such structures are called retia mirabilia, and they are strategically located in the ciculatory system so that the blood both to and from some tissue or organ must pass through them. The blood flows in opposite directions in the two sets of vessels, and thus makes possible a counter-current exchange of any agent that can diffuse across the thin boundary between them. The most studied retia are those of the swimbladder which, in deep-water fish, may contain oxygen at a pressure of 200 atm. A rete restricts the loss of some agent, oxygen or heat, which would otherwise be carried away in the venous blood; it imposes a high resistance to its dispersion. Retia do not greatly restrict blood flow for the resistances of the many retial vessels are resistances in parallel (Chapter 1). Neither do they impose an infinitely high resistance to the loss of the agent to be conserved, but their efficiency is impressive. Scholander (1957) calculated that across a rete with capillaries 1 cm long, of a swimbladder containing oxygen at a pressure of 200 atm, the P_{O_2} of the outflowing blood is reduced 3000 times. Translating this in terms of heat, it is the equivalent of an exchanger in which iced water goes in at one end, and boiling water at the other, yet the transfer of heat between the two is complete to within 0.0001 °C.

The three significant dimensions of a rete are the thickness of the wall, the length of the vessels, and the rate of flow through these; altering any one will favour or impede the concentration of the agent.

Retia that concentrate oxygen and nitrogen
The retia of the swimbladder

The swimbladder is primarily a hydrostatic organ which renders some teleosts neutrally buoyant and bottom-living fish which lack a swimbladder are commonly 5–6% denser than sea water, depending on the proportion of fat and bone in their bodies. The total pressure of the gases within their swimbladders must approximately equal the hydrostatic pressure surrounding them and fish with swimbladders are common at depths of 200–1000 m. These depths roughly correspond to pressures of 20–100 atm. The swimbladder gases of deep-water fish have oxygen as the chief component and a typical analysis is that of Schloesing and Richard (1898), from a fish caught at 900 m: O_2 75%, N_2 20.5%, CO_2 3.1%, argon 0.4%. Shallow-water fish have swimbladder gases with a composition similar to air, with nitrogen as the dominant gas. Perch, after many weeks in a shallow aquarium, had 80% of N_2 in their swimbladder gases. Carbon dioxide can be abundant, and Wittenberg *et al.* (1964) report that in blue fish it constituted 37%. As fish from greater depths are sampled, oxygen becomes increasingly dominant and the ability of the gas gland–retial system to concentrate it, the more remarkable. The ocean can be thought of as a column of molecular interstices extending from the surface to the bottom, through which the atmospheric gases diffuse. Hence, close to the bottom, the P_{O_2} and P_{N_2} of the water, and of the arterialized blood, will be as they are at the surface, 0.2 and 0.8 atm, respectively. A fish with a swimbladder of P_{O_2} of 100 atm and a P_{N_2} of 25 atm must thus have an ability to concentrate oxygen 500 times and nitrogen 31 times (Steen 1970). How, we may ask, are such gas concentrations, so far above those prevailing in any other organ or tissue, produced?

The swimbladder of an eel (Figure 28*A*) bears, on its external surface, a pair of retia; they are supplied with arterial blood from the pre-retial

artery, a branch from the dorsal aorta, and venous blood from them
flows, via the postretial veins, into the hepatic portal vein (Steen 1963).
The rete is composed of arterial and venous vessels of capillary dimensions. Blood from the arterial capillaries at the bladder pole of the rete
(Figure 28B) passes, in a post-retial artery, to supply the gas gland on the
inner wall of the bladder. Blood from this flows, via a pre-retial vein, to
the venous capillaries of the rete. Such a structure is termed a bipolar rete;
in many species the capillaries of the rete run directly into the gas gland,
constituting the so called unipolar rete. In the eel, a rete of average size,
with a cross sectional area of 5 mm^2 and a volume of 21 mm^3, would have
some 34 000 arterial and 22 000 venous capillaries. Arterial capillaries are
9–10 μm in diameter, venous capillaries 11–13 μm; the diffusion distance
from arterial to venous capillary is about 1 μm. The capillaries are rather
uniformly 4 mm long. The surface areas of the two sets of capillaries are
30 cm^2 and are equal, for the smaller number of venous capillaries is
compensated by their larger diameter (Stray-Pedersen and Nicolaysen
1975).

Ultrastructural studies show that the two sets of capillaries differ in
other respects. The walls of both consist primarily of an endothelial layer
with a basal lamina but some interstitial tissue of collagenous and elastic
fibres is present in the angles between the vessels. The walls of the venous
capillaries have fenestrae, 20–80 nm in diameter, covered with a 5 nm

Figure 28. Swimbladder retia of the eel. *A*, Ventral view. *B*, Diagram
of a bipolar rete. Redrawn after Steen 1970.

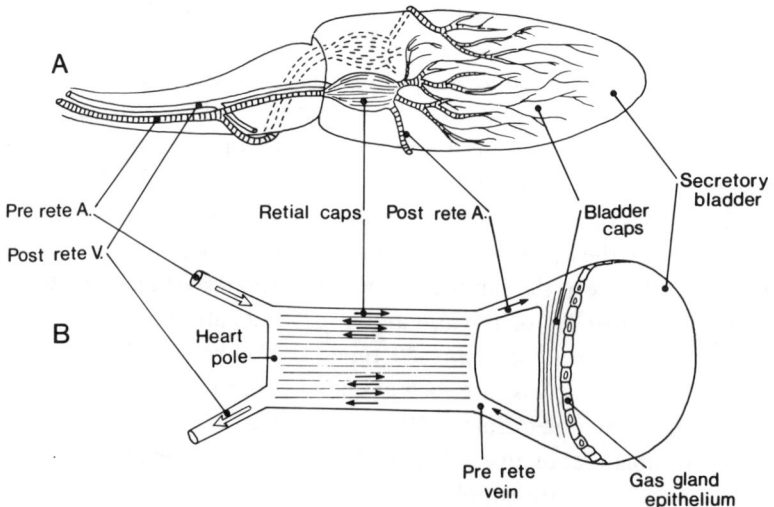

diaphragm and these fenestrae are most common where the walls are thinnest. Venous endothelial cells are considerably lower than those of arterial capillaries which have a cuboidal appearance and a rather vacuolated cytoplasm, showing pinocytic vesicles and branching tubular invaginations (Bendayan *et al.* 1974).

The capillaries of the gas gland lie below an epithelium which has the ability to produce lactic acid by glycolysis even at oxygen pressures of 100 atm (D'Aoust 1970). Lactic acid enters the arterialized blood and causes oxygen to be liberated from it by three mechanisms. The increase in proton concentration decreases the affinity of the blood for oxygen: the Bohr effect. The increase in acidity also diminishes the O_2 capacity of the blood: this is the Root effect (Chapter 4). The presence of other ions in the blood, notably H^+ and lactate, diminishes the solubility of O_2 in the blood: this is the salting-out effect. That the Bohr and Root effects are separate phenomena is now disputed (Chapter 4). Nevertheless, acidification of the blood, by the outpouring of lactic acid into it, causes a reduction of both its oxygen affinity and its oxygen capacity. The net effect is that oxygen passes from combination into solution and this raises its P_{O_2} as it leaves the gas gland capillaries and enters the pre-retial vein and the venous capillaries (Figure 2B). Venous blood with a high P_{O_2} flows along the venous capillary blood. This blood will already have gained some oxygen by diffusion but the venous P_{O_2} at this point is sufficiently high for more to enter into solution. As blood passes further along the venous capillary towards the heart pole, its P_{O_2} falls further as its contained oxygen diffuses across the thin capillary walls into the arterial blood, but further diffusion is still possible as it is now partnered by arterial blood with a lower P_{O_2}. A gradient of oxygen pressure exists along the whole length of the retial capillary. The capillary walls are permeable to H^+ and these ions too diffuse from the venous to the arterial blood. This loss of H^+ from the blood in the venous capillaries does not lead to an immediate loading of oxygen onto the haemoglobin there. If this were to occur it would lower the venous P_{O_2} and bring diffusion to a halt. There is a significant difference in the time constants for the on-loading of oxygen (10–20 sec) and the off-loading (50 msec) which minimizes this (Berg and Steen 1968, Forster and Steen 1969). By the time the blood in the venous capillaries has responded to the reduction in proton concentration it has passed out of the rete into the hepatic portal vein. The concentration of oxygen which can be attained in the bladder will thus be related to the rate of flow of blood through the rete. Sund (1977) has mathematically modelled the system using established data and predicted that there is an optimal rate of blood flow through the rete which will achieve the maximal rate of gas production, and flows above or below this will both result in lower rates. His model also predicted that the gradient of P_{O_2} will peak at a position towards but not at the bladder pole of the arterial capillary, and that this position will shift towards the heart pole as the P_{O_2} in the arterial blood rises. Steen and Sund (1977) have measured blood P_{O_2} along the length of the rete with O_2-sensitive microelectrodes and shown that this does indeed occur.

The concentration of nitrogen in the swimbladder must depend on the salting-out effect. The addition of lactic acid to the plasma reduces the solubility of nitrogen in it which releases some of the gas present in physical solution. This must also be the sole mechanism for further oxygen concentration when there are such high pressures of oxygen in the bladder that the haemoglobin is fully saturated even when acidified to the maximum extent. Noble *et al.* (1975) report that the blood of the benthic fish *Antimora rostrata*, at the lowest pH to which it might attain, would be fully saturated at the 300 atm pressure in which it lives, and that salting-out remains the only means we know of, by which oxygen might be further displaced into the swimbladder.

The mechanism outlined above is an example of a counter-current multiplier system, and it differs from those involved in the concentration of heat, discussed in the next section, which are purely counter-current systems. They can only retain the heat generated in the muscle; in the swimbladder rete, the repeated recirculation of oxygen and protons back through the rete and gas gland leads to successive rises in the P_{O_2} of blood in the bladder capillaries and the eventual liberation of gaseous oxygen into the bladder.

The removal of gas from the bladder

Some fish, such as the salmonids, carp and eels, have a pneumatic duct through which air can be released to the surrounding water. In the majority of fish possessing a swimbladder (Figure 29*A*, *B*), it is totally closed and the gases it contains are removed in solution in the blood; this is the physoclist condition and is found, for example, in all the perciform fish, a suborder containing over a hundred families. The bladder comes to be divided into two parts, each with a separate circulation. A rostral secretory region is lined with a thin mucosa beneath which is a network of capillaries supplied by blood with a high P_{O_2} from the retia. The blood flows to these vessels from the coeliacomesenteric artery, and is returned to a branch of the hepatic portal vein; it can be compared with part of the gut in this regard. A caudal reabsorbent region also has a capillary network beneath its mucosa, but this is supplied, in its central region, directly from the dorsal aorta, and in its more lateral parts, by intercostal arteries (Ross 1979a). Blood from this enters the posterior cardinal veins (Fänge 1983). The reabsorbent capillary net is richly supplied through muscular contractile arterioles which are known to be innervated (Figure 28*B*, At.) (Fänge 1976). In many of the advanced physoclists, the reabsorbent area is concentrated into an oval patch surrounded by a circular sphincter, formed from a concentration of the muscularis mucosae. This also forms radial bands of muscle and the oval area can be expanded by contraction of these, or restricted, by contraction of the sphincter. In others, such as the wrasses (*Labrus, Crenilabrus*) (Figure 29*A*), an incomplete diaphragm separates the secretory and reabsorbent parts of the swimbladder.

The volume of air within the bladder, and hence the buoyancy of the fish, is regulated by the autonomic nervous system, and its control will be described in Chapter 9. We can note here that a deflatory reflex is recognized, when secretion is minimized and reabsorption is maximal,

Figure 29. The reabsorbent mechanism of the physoclist swimbladder *A*, Bladder of goldsinny wrasse in the secretory phase. *B*, Diagram of circulation to bladder in reabsorbent phase. At. arterioles, A.V.N. artery, vein and nerve to gas gland and diaphragm, D. diaphragm, F.Cm.A. from coeliacomesenteric artery, F.D.A. from dorsal aorta, G.Gl. gas gland, T.H.V. to hepatic portal vein, T.P.C.V. to posterior cardinal vein. Rb.R. reabsorbent region, Rt. rete, Sc.R. secretory region. Redrawn after Fänge 1966, 1976.

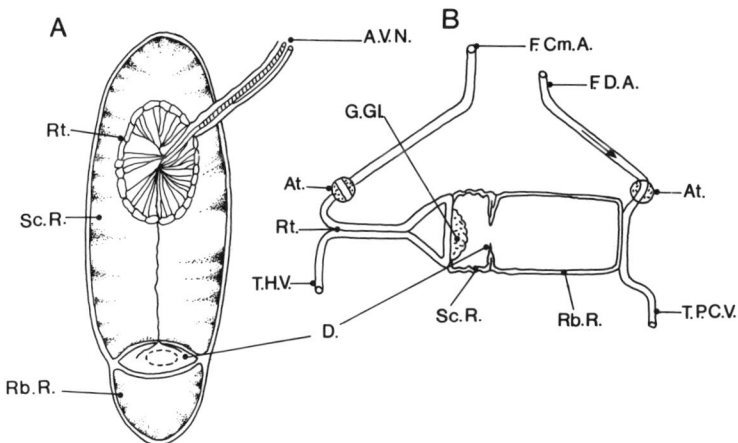

and an inflatory reflex when the balance of the two processes is reversed. The inflatory reflex is brought into action if the bladder is emptied artificially. Under natural conditions the position of the diaphragm, mentioned above, moves rostrally as deflation occurs so that a maximal area of reabsorptive mucosa is exposed (Figure 29*B*). During inflation the diaphragm moves posteriorly, extending the secretory mucosa over most of the bladder wall (Figure 29*A*) (Fänge 1976, Fänge *et al.* 1976).

The choroidal rete of the eye

The retina of teleost fish, like that of higher vertebrates, has a high respiratory rate yet, except in the eel, it lacks both superficial and interretinal blood vessels and is indeed, totally avascular. Oxygen is utilized via an active TCA cycle and is obtained by diffusion from a capillary network in the choroidal layer behind. Fairbanks *et al.* (1969) report that in the trout, the retina has a P_{O_2} of 445 ± 68.5 mm Hg; this is 20 times that of the arterial blood and 3.5 times that of the ambient water. Wittenberg and Wittenberg (1962) list other species with similar high levels of ocular P_{O_2}, e.g. the remora (*Echeneis naverales*) 435–1320 mm Hg, blue fish 240–820 mm Hg and puffer (*Sphaeroides maculatus*) 365–575 mm Hg.

In 1806, Albers, a Göttingen ophthalmologist, described a horse-shoe shaped vascular structure in the eye of the bowfin and recognized that it

Figure 30. The choroidal rete of the trout. The sclera and choroid layer of the eyeball are shown in surface view and in cross section. A.C.R. arterial capillaries of rete, C.C. choriocapillaris, Ch. choroid layer, L.B. lentiform body, O.A. ophthalmic artery, O.N. optic nerve, O.P.S. ophthalmic venous sinus, R.A. retinal artery, R.C.R. right half of choroid rete, R.V. retinal vein, Sc. sclera. Redrawn after Barnett 1951.

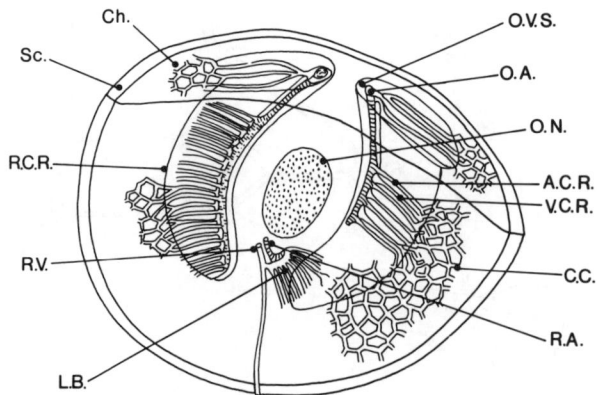

was a rete. Barnett (1951) has described the choroidal retia of nine species of teleosts. In the trout (Figure 30) the two arms of the horse-shoe surround the optic nerve and comprise curved branches of the ophthalmic artery each of which lies within the wider ophthalmic venous sinus. From these, retial vessels pass out towards the periphery; they are of capillary dimensions and are closely packed together like those of the swimbladder rete. The choroidal rete is unipolar and its arterial and venous capillaries give rise directly to the capillary network of the choroid. In the remora and blue fish the rete is so large that it causes the external surface of the eyeball to bulge. Below the optic nerve lies a smaller second rete, the lentiform body, which is separately supplied with blood from the retinal artery. Capillaries from the lentiform body supply a smaller area of the choroid below the optic nerve. The mechanism of oxygen concentration is presumed to involve an acidification of the blood as it flows through the choroidal capillaries, and an increase in its P_{O_2} as oxygen is displaced from combination into solution by a combination of the Bohr and Root effects. Like the swimbladder rete it is a counter-current multiplier system. The source of the protons is uncertain; they may derive from lactic acid or from CO_2. The mechanism is known to be dependent on carbonic anhydrase which suggests that H_2CO_3, derived perhaps from the actively respiring retina, is the source of the protons.

Blood is supplied to the choroidal rete from the ophthalmic artery which rises from the pseudobranch. This structure lies beneath the inner lining of the operculum and is believed to be a vestigial gill; it contains acidophilic cells rich in carbonic anhydrase. Its role in the oxygen-concentrating mechanism is uncertain; earlier studies reported that pseudobranchectomy lowered the oxygen concentration of the eye. However, the trout retina too has a high level of the enzyme, and is known to be able to carry on glycolysis in the presence of hyperbaric oxygen. There is also evidence from *in vitro* studies of the trout retina, that acetazolamide directly inhibits its oxygen-concentrating mechanism (Hoffert and Fromm 1972, 1973). We noted in Chapter 4 that acetazolamide blocks the action of carbonic anhydrase and its injection lowers the P_{O_2} of the trout retina from 400 mm Hg to the arterial level within 2 min of injection. This in turn leads to the loss of the B wave of the electroretinogram, 1 min later (Fonner *et al.* 1973). Maintaining ocular P_{O_2} at only arterial levels is not sufficient to sustain retinal function. Such studies give a graphic indication of how active the concentrating mechanism and the flow of blood to it is, and how important it must be in the lives of predaceous fish that need sharp vision.

Ingermann and Terwilliger (1982) examined 15 species of teleost, including the Dover sole, the smooth sculpin (*Artedius lateralis*) and the black prickleback (*Xiphister atropurpureus*), all of which lack a swimbladder, yet have bloods with a pronounced Root effect. All, however, have choroidal retia. On the basis of this the authors challenge the supposition that Root effect haemoglobins evolved in association with the swimbladder rete, and propose as an alternative that they first appeared in response to the need to raise the oxygen tension in the eye of sight-oriented predaceous fish. Only subsequently, they suggest, were the Root effect haemoglobins additionally harnessed to the task of filling the swimbladder. A similar conclusion has been drawn by Dafré and Danilo (1989) from a study of 46 species of Brazilian marine fish.

Retia that concentrate heat

The responses of the circulatory system to temperature change in fish lacking retia

Fish can undergo great changes of ambient temperature in the course of a year. Winter flounder of Newfoundland may experience a temperature range of $-1.5\,°C$ to $+16\,°C$ (Fletcher 1977) and such changes have diverse effects on different features of the circulatory system. Increase of temperature reduces the viscosity of the blood and thus lessens the work that the heart has to perform to achieve a given flow. It diminishes the solubility of oxygen in the plasma, which may be a significant proportion of the oxygen content of the blood of some fish. It increases the rate of enzyme reactions and thus elevates both the oxygen demands of the tissues, and the rate at which the heart can perform work. Both ventilation and heart rates of trout increase by about 34% between 15 and 24 °C, but stroke volumes change little and the increase is small compared with the nearly threefold increase in oxygen consumption (Heath and Hughes 1973).

Circulatory responses to changes in temperature can be altered by acclimation. This is a process which usually takes days or weeks, but can sometimes be seen to commence in a few hours, and it tends to off-set the change initially caused by the alteration of temperature. As a result of temperature acclimation such as variables as the oxygen consumption of a fish, its heart rate and its cardiac output will, at a given temperature, be greater in winter than in summer. The metabolic rate at 5 °C may be as high in winter as it is at 15 °C in summer.

Acclimation to temperature

Acclimation can take many forms; it can be seen as an increase or decrease in tissue elements. Trout kept at 5 °C increase the amount of muscle in the ventricle and thereby partly off-set the reduction in myocardial power output observed acutely (Graham and Farrell 1989). Johnston (1982) reported that carp skeletal muscle has more capillaries in winter than in summer, reflecting perhaps the fact that temperature is a term in Fick's equation of diffusion. The skeletal muscle of striped bass (*Morone saxatilis*) acclimated to either 5 °C or 25 °C shows changes in mitochondrial abundance; the fraction of cell volume occupied by mitochondria in red muscle increases from 0.28 at 25 °C to 0.45 at 5 °C (Egginton and Sidell 1989). Intrinsic regulatory responses may be important; cold acclimated (5 °C) perfused isolated hearts of trout have a lower heart rate than they do at 15 °C but the stroke volume is the same or greater due to the longer time for filling (Graham and Farrell 1989). Other acclimatory changes involve the autonomic nervous system (Chapter 9). In the eel the increase in heart rate caused by temperature rise is in part off-set by an increase in vagal tone (Seibert 1979); in other species the change may involve a reduction of adrenergic stimulation. The trout heart at 5 °C has a greater β adrenergic sensitivity and is more dependent on a low level of adrenaline for its continued beat than it is at 15 °C (Graham and Farrell 1989). In terms of its maximum cardiac output the cold acclimated perfused trout heart has a 73% compensation of the reduction that is seen acutely ($Q_{10} = 1.27$); this is achieved by a 43% compensation for rate ($Q_{10} = 1.55$) and a 30% increase in stroke volume (Graham and Farrell 1989).

Nevertheless, whilst some of the changes acutely imposed on the circulatory system by temperature change can be compensated by acclimation, others are little affected. In the trout, acclimation to 5 °C, 10 °C and 15 °C had no significant effect on the concentration of ATP in the blood, nor on its oxygen affinity (Weber *et al.* 1976). The advent of homoiothermy, in the higher vertebrates, must have had clear advantages. It not only enabled an elevated and constant temperature to be maintained throughout much of the body: it also permitted the closer dovetailing together of the rates of many of the different cellular and organismal subsystems that go to make up the circulatory system.

The body temperatures of the majority of fish are, at most, no more than 2 °C above ambient (Stevens 1973). The heat capacity of water is 3000 times that of air and 80–90% of a pulse of heat injected into the aorta is lost during its passage through the gills (Erskine and Spotila 1977, Stevens and Sutterlin 1976). The metabolic advantage that might accrue if the heat generated by a muscle's metabolism could be retained, is thus not normally an option available to a fish because of the 'in series' arrangement of the gills and the tissues. In two quite unrelated groups of fish, the tunas and their allies (Suborder Scombroidei), and the mackerel sharks, elasmobranchs of the family Lamnidae, comprising such species as the mako (*Isurus oxyrinchus*), the porbeagle (*Lamna nasus*) and the white pointer (*Carcharodon carcharias*), the red myotomal muscles can be maintained well above water temperature, T_w, through the presence of muscle retia.

The muscle retia of tuna

We noted in the Chapter 6 that the red myotomal muscle (Figure 27) is in most fish located as a thin lateral fillet just beneath the skin, and any heat it may generate is likely to be lost directly through this into the water. In the sea raven, 70–90% of the metabolic heat is lost through the skin of the body and fins (Stevens and Sutterlin 1976). The most obvious change in myotomal structure in tuna is a shift in the position of the red muscle from a peripheral position to a central one, where it is thermally insulated by the surrounding mass of white muscle.

The skipjack tuna (Stevens *et al.* 1974) has a large median rete (Figure 31) above the dorsal aorta and posterior cardinal vein. In a 1.9 kg fish it is

Figure 31. The muscle retia of the skipjack tuna. A. + V.cap.R. arterial and venous capillaries of rete, D.A. dorsal aorta, K. kidney, P.C.V. posterior cardinal vein, S.A. segmental artery carrying warm blood to muscle, S.M. segmental muscles, Sp. spine, S.V. segmental veins carrying warm blood from muscles. Redrawn after Steven *et al.* 1974.

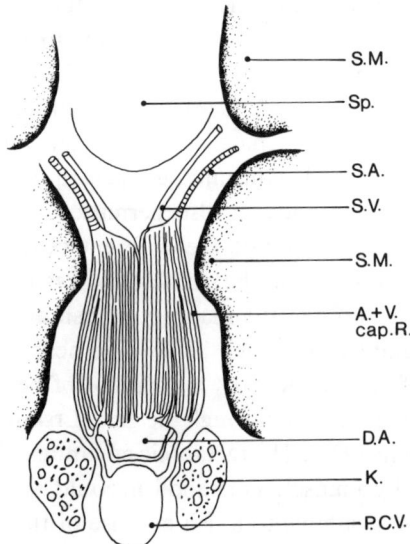

about 7 mm wide and its vessels are about 1 cm long. It is a single median structure crowded in between the inner margins of the myotomes, the kidney and the spinal column. From the dorsal aorta, numerous small arteries carry blood to the retial arterioles. The outflows from these coalesce to form segmental arteries, which carry the warmed blood to the red muscle. Blood from this passes through the retial venules and enters the single median posterior cardinal vein.

The retial arterial vessels are the diameter of arterioles, 36 μm, and have a smooth muscle layer which lacks innervation. Venous vessels are larger in diameter than arterial vessels, 84 μm, and have thinner walls. Alternating arterioles and venules are packed together in a rectangular array, with virtually no other tissue between them. The combined walls have a thickness of 10 μm. This is 10 times that in the swimbladder rete. Stevens *et al.* (1974) calculate that, in total, the rete of a 1.9 kg skipjack comprises 133 721 arterioles and 120 652 venules with internal surface areas, respectively, of 1350 and 2858 cm^2 and that it contains 5.2% of the blood volume.

From the cardiac output, (154.4 ml min^{-1}) and assuming that 60% of this goes through the rete, they calculated that this volume, 92.6 ml min^{-1}, flows with a velocity of 1.15 cm sec^{-1} through the narrow arterioles and 0.23 cm sec^{-1} through the wider venules. The blood is in the rete for about a second and has its velocity reduced to about $\frac{1}{80}$ that in the main vessels leading to and from it. This slow flow provides the time for heat to pass from the outgoing venous blood back into the arterial vessels. The total diffusion area, in relation to the volume of the rete, is 187 cm^2 cm^{-3}. This is an order of magnitude smaller than the comparable figure of the eel swimbladder rete, of 1700 cm^2 cm^{-3}. Stevens *et al.* (1974) calculate that these data are such as to favour heat transfer but not to allow the respiratory gases to exchange between the arterial and venous blood.

Various measurements of muscle temperatures in wild skipjack tuna (Barrett and Hester 1964, Stevens and Fry 1971) confirm that they are appreciably warmer than the water. In 20 °C water, skipjack red muscle may be 8 °C warmer. The excess temperature T_X, is greater in colder water and increases with the weight, W. From many measurements Stevens and Neill (1978) derived the following equation:

$$T_X = 1.45 \; W^{0.58}$$

When skipjack tuna are kept in captivity, the magnitude of T_X gradually falls, although the physiological changes that bring this about are not understood. Exercising captive skipjack increases T_X a little, but not to the values seen in wild specimens.

Big-eye tuna (*Thunnus obesus*) and bluefin tuna (*T. thynnus*) have a totally different arrangement of muscle retia. The red muscle retains its medial position but the blood supply to it is shifted to a more lateral position thereby providing space for the insertion of a rete. The dorsal aorta and posterior cardinal vein, which in skipjack, as in other fish, are the main vessels afferent and efferent to the myotomes, are here quite small. The lateral cutaneous vein, in most fish a narrow vessel draining the skin, is large and double (Figure 32*A*) and each branch is paralleled by a corresponding artery, a new vessel derived from the dorsal aorta at the

Figure 32. The muscle retia of the big-eye tuna *Thunnus obesus*. *A*, Semidiagrammatic figure of retia, red muscles and their vascular supply. *B*, Diagram of retial counter current flow in a unipolar rete. C.A. cutaneous arteries to rete, D.A. dorsal aorta (small), L.C.V. lateral cutaneous veins from rete, P.C.V. posterior cardinal vein (small), R.A. retial arteries, R.M. red muscle, Rt. upper and lower retia, Sp. spine. Redrawn partly after Carey *et al.* 1971.

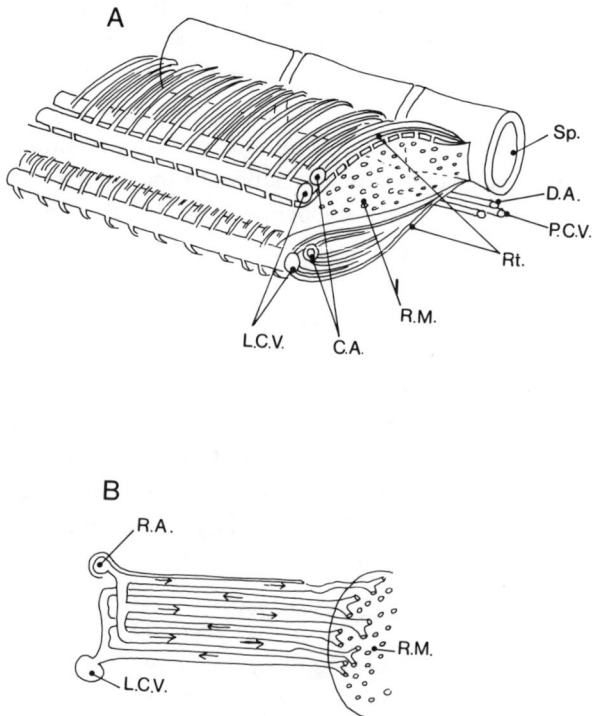

level of the epibranchial arteries (Kishinouye 1923, Carey *et al.* 1971). From these lateral vessels, numerous parallel arteries and veins pass centrally towards the spine and supply a mass of retial vessels both above and below the red muscle. These vessels are of arteriolar dimensions, 100–200 μm in diameter, and in a large bluefin tuna form 1 cm thick slabs of retial tissue both above and below the red muscle fillet, extending most of the length of the fish. From the rete, vessels of arteriolar dimensions pass directly into the red muscle; this is therefore a unipolar rete (Figure 32*B*). Temperature measurements of the red muscle of bluefin tuna, made by telemetry (Carey and Lawson 1973), show that it is commonly maintained 8–10 °C above T_w. In large bluefin tuna in cold water (7 °C), it may be 19 °C warmer (Figure 33*A*).

Figure 33. Tissue and organ temperatures in bluefin tuna and mako shark. *A*, temperature of red muscle, eye and brain of bluefin. *B*, temperature of red muscle of mako. In *A* and *B* tissue temperature in a poikilothermous fish would fall on the dashed line of body temperature = water temperature. *C*, Stomach temperature (upper trace) and water temperature (lower trace) in a bluefin. Abscissa = time in days after release. Arrows indicate large feeds. Redrawn, *A* after Scott-Linthicum and Carey 1972, *B* after Carey and Teal 1969b, *C* after Carey *et al.* 1984.

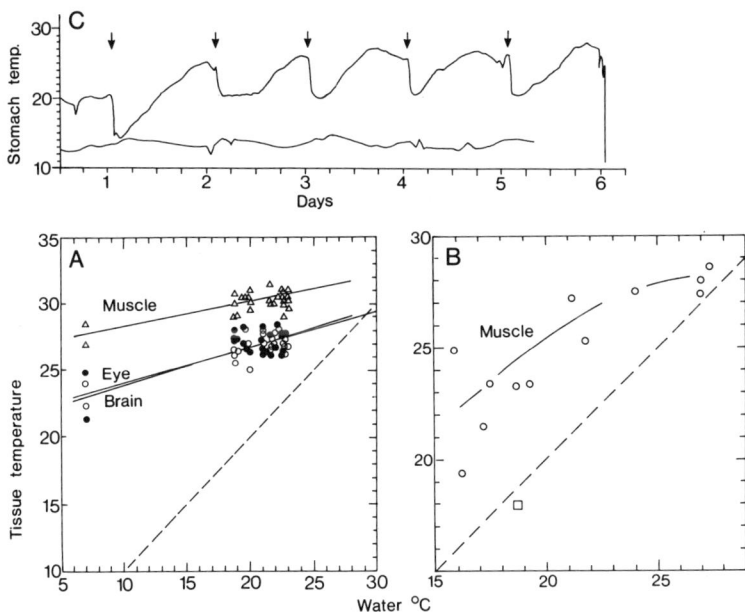

Telemetric measurements, made as these fish swim through the thermocline, suggest that they can thermoregulate their red muscle, for its temperature falls only 6 °C between water temperatures of 20 and 35 °C (Carey and Teal 1969a). To some extent this may be due to the fact that, in warmer water, tuna tend to moderate their activity as their muscle temperature rises towards a critical limit; moreover, their large size provides a great thermal inertia. Nevertheless there appears also to be some, as yet unknown, way by which in warm water, the retia can be bypassed and their heat-conserving ability degraded. Such a mechanism is thought also to exist in the retia associated with visceral warming in the mackerel sharks (Carey *et al.* 1981).

Muscle retia in the mackerel sharks

It is surprising how closely similar are the retia associated with the red muscle in the mackerel sharks, compared with those of a bluefin tuna. In the porbeagle shark (Burne 1923), the red muscle is intucked to form a lateral fillet, disposed more centrally than in tuna. A single rete on each side is associated with it; it lies lateral to the red muscle and is supplied by a prominent artery and vein which run down beneath the skin, close to the lateral line. The vein, once again, is the enlarged lateral cutaneous vein; the artery is a new vessel which arises from the fourth and fifth epibranchial arteries. The reduction of the normal dorsal aortic supply has proceeded even further than in the bluefin, and the epibranchial arteries beyond the origin of the lateral artery are almost without a lumen. The posterior cardinal veins are represented only by a single median vessel which disappears altogether at the level of the first sympathetic ganglion.

In the mako, the retial vessels form a single slab of tissue but in porbeagle and white pointer sharks they are diffused amongst the red muscle fibres, and appear as clusters of three, arranged as trios of vessels: vein, artery, vein. Carey and Teal (1969b) report that the warmest muscle in the porbeagle may be 20 °C above T_w when this is 11 °C. Figure 33*B* shows that the curve of muscle T against water T_w in mako sharks is less steep in warmer waters; the warmest portion of the muscle mass varies only 5 °C over a 10–30 °C range of water temperatures.

The functional advantage of these elevated red muscle temperatures is uncertain. Tunas are the fastest swimmers of all fish, and a number of processes involved in muscle contraction are temperature sensitive. The time that elapses between electrical stimulation of fish muscle and its peak of contraction is shortened by an increase in temperature as is the dispersal of muscle lactate following a bout of great activity. Red muscle contains myoglobin and few animals have redder muscles than tuna (Stevens and Carey 1981). Myoglobin facilitates the movement of oxygen from the cell membrane to the mitochondria; it loads on oxygen at the membrane and, by translational movements within the cell, carries it to the mitochondria where it is off-loaded. This process of facilitated diffusion is also temperature sensitive, and is increased some 40%, by a 10 °C rise in temperature.

Retia that concentrate heat in the digestive organs

In 1835, Eschricht and Müller described retia that occur on the dorsal surface of the liver in the bluefin tuna and Carey *et al.* (1984) have added further details of these structures. The hepatic portal vein, prior to its entry into the liver, breaks up into numerous retial vessels, grouped into three masses, and these are partnered by retial arterioles derived from the coeliac artery. The portal vein blood, warmed by the aerobic metabolism of digestion, and by the heat derived from the hydrolysis of the fat and protein in the food, thus has the opportunity to return this heat to the incoming arterial blood. The viscera are insulated from conductive losses by the swimbladder above and the mass of fatty red muscles at the sides. Carey *et al.* (1984) noted that the blood passing out of the rete, into the liver, was at water temperature, as was the arterial blood. Their acoustic telemetry studies showed (Figure 32C) that the stomach contents were initially cooled as the food entered, but that the stomach, pyloric caecae and intestine then gradually increased in temperature by 10–15 °C over a period of 12–20 h. Captive bluefin tuna eat some 20 kg of herring and mackerel at a feed, and Carey *et al.* (1984) calculated that the hydrolysis of the 3 kg of fat and 4 kg of protein in this should yield 130 kcal (544 kJoule), which should raise the visceral temperature 5 °C. The remaining increase is presumably due to the enhanced aerobic metabolism associated with digestion. Stevens and McLeese (1984) have investigated the advantages that accrue from warming the viscera, and focussed their attention on the digestive enzymes, trypsin and chymotrypsin, as hydrolysis is commonly the rate-limiting step in the release and absorption of amino acids. They showed that their specific activity and maximal reaction velocity increases with a Q_{10} of around 2 within the physiological temperature range. Such an increase permits the tuna to digest protein in about a third of the time, and thus to process a greater quantity of food each day.

The mackerel sharks also have paired suprahepatic retia and can retain heat in their digestive organs but the complexity of their retia is less than in tuna, and their blood supply is quite different. Burne (1923) and Carey *et al.* (1981) have described the retia of the porbeagle and mako sharks. The two retia are not, as in the bluefin tuna, an array of alternating arterial and venous vessels but are, rather, two skeins of arterioles suspended, side by side (Figure 34), in a venous sinus. Carey *et al.* (1981) give the dimensions of a rete from an 88 kg porbeagle shark as 9 cm long by 18 cm wide; it tapered from 3 cm thick at the anterior end to 1.5 cm at the posterior end. This retial tissue contained 41 arterioles per mm^3, each with an average diameter of 130 μm. The venous sinus containing the rete lies on the underside of the oesophagus and is derived from the hepatic vein. The arterial supply to the rete is derived from continuations of hypobranchial vessels which also give rise to the coronary arteries. The contributions from the second, third and fourth branchial loops combine to form a large vessel on each side, which passes backwards to divide into the mass of arterioles of the rete. The blood, following its passage through the retial arterioles, passes into a number of large arteries, which supply the stomach and intestine. The coeliac and mesenteric arteries branching from the dorsal aorta, which normally would supply these organs, are much reduced.

The direction of blood flow in the arterioles must be away from the heart towards the viscera, and the direction of blood flow through the

Figure 34. The suprahepatic rete of the Porbeagle shark. A. atrium, At. arteries from rete to stomach, intestine and other viscera, C.A. coronary artery, E.B.V. efferent branchial loop vessels around gill slits 1–5, H.S. hepatic sinus, H.V. hepatic vein, L. liver, Oe. oesophagus, S. sinus venosus, S.H.R. suprahepatic rete, V. ventricle. Redrawn partly after Burne 1923.

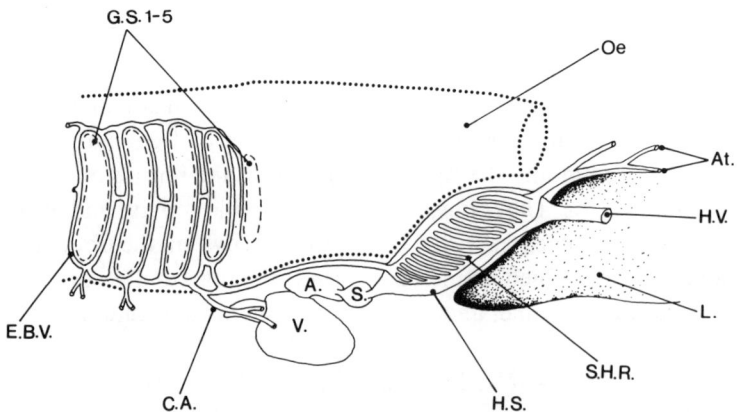

large sinus of the hepatic vein is opposite to this; in this sense the two flows are counter-current, the one to the other. But the warm venous blood is not channelled through specific venules; it merely passes through irregular spaces between the arterioles. The structure appears to be less highly evolved than the retia described in the previous section. In addition, there is a median channel of larger diameter, which extends forwards between the two retia, and allows warmed blood to bypass the heat exchange mechanism. Carey *et al.* (1981) note that it has smooth muscle in its wall, and it may be able to regulate retial function. Despite the simple structure of these suprahepatic retia, heat is conserved, and Carey *et al.* (1981) report that the stomachs of mako sharks are 6–8 °C warmer than the surrounding water.

Retia that concentrate heat in the brain and eye

On each side of the head of bluefin tuna, a carotid rete occurs, dorsal to the first efferent branchial artery and ventral to the prootic bone of the skull to which it is attached. Its structure is more complex than the suprahepatic retia of the shark, described above, yet less perfected than the closely packed mass of vessels that form the muscle rete in the skipjack tuna. It consists of separated arterioles, each surrounded by a ring of 4–9 venules. The venules will contain blood warmed by the metabolism of the retina and brain. The groups of vessels occur scattered in an extensive mass of pigmented intervascular tissue (Scott-Linthicum and Carey 1972). The arterioles are 80–120 μm in diameter and have 20 μm thick walls, rich in smooth muscle. The venules are 40–150 μm in diameter. Blood from the secondary lamellae of the first gill passes to the rete mainly via the common carotid artery and smaller contributions occur from the afferent pseudobranchial and opercular arteries; this blood will be at the temperature of the water. Blood warmed by its passage through the rete leaves via the external carotid artery, enters the skull and supplies the brain. Just before it enters the skull it gives off the ocular artery which carries warm blood to the vascular network of the retina.

Scott-Linthicum and Carey (1972) have measured brain and eye temperatures in bluefin tuna (Figure 32*A*) and report that in large specimens (181–408 kg) these tissues are some 7 °C warmer than the water. The equations for the regression lines of brain and eye temperature with respect to water temperature T_w are:

brain temperature $\quad 0.28 \ T_w + 21.1$ °C

eye temperature $\quad 0.30 \ T_w + 21.1$ °C

The similarity of these equations supports the evidence of their vascular anatomy, i.e. that brain and eye derive their warm blood from the same rete. The posterior half of the eye is generally warmer than the aqueous humour and one bluefin from 7 °C water had a retinal temperature of 22.4 °C and an aqueous humour temperature of 12 °C. This gradient presumably reflects the many small vessels that spread out behind the retina. Other genera and species of tuna have carotid retia and Scott-Linthicum and Carey (1972) have recorded brain and eye temperatures 4 °C or more above T_w in albacore, big-eye and black skipjack (*Euthynnus alleteratus*), as well as in bluefin tuna, whereas other very active fish which lack a carotid rete, such as mackerel and blue fish, have brain and eye temperatures within 0.3 °C of T_w.

Mako and porbeagle sharks also have an orbital rete in the venous sinus that surrounds the back of the orbit (Block and Carey 1985) and temperatures in the brain and eye are 5 °C higher than the surrounding water; brain and eye derive their blood from the same rete. In these fish the arrangement recalls that of the suprahepatic rete in that there are not separate parallel venous channels, for the arterial retial vessels are washed by the slowly moving sinus blood.

Yet other families have evolved the capacity to warm the eye and brain, by coupling the carotid rete with a mass of brown tissue; less is, at present, known about the structure of these organs. In the swordfish (Family Xiphiidae) and in various species of marlin such as the blue, and the striped *Tetrapterus audax* (Family Istiophoridae), the carotid rete is followed by a mass of brown tissue the cells of which are rich in mitochondria. This thermogenic tissue has features in common with the brown fat of higher vertebrates, but the heat it generates is prevented from dispersing to the rest of the body by the carotid rete and is thus conserved to raise the metabolic rate of the eye and brain (Carey 1980, 1982).

Fish are unique amongst the lower vertebrates in that these two groups, the tunas and the mackerel sharks, have repeatedly attained a regionalized warm-bloodedness by means of retia. Moreover, the extent of the ability to conserve metabolic heat, in large species like bluefin tuna, can apparently be regulated depending on water temperature, and whilst the mechanism of this is not yet understood, it is presumably localized at the level of the individual retia. It thus contrasts with the overall warm-bloodedness of the birds and mammals, regulated from nervous centres in the hypothalamus.

It should be emphasized that tuna and mackerel sharks are in no way related. Elasmobranchs have been in existence since the middle Devonian;

then as now, some of them were specialized carnivores of large size. Teleosts do not appear until the Triassic (Stahl 1974), and the tunas (Family Thunnidae) are amongst the most specialized of them. The feature that the various fish discussed in this chapter have in common, apart from their ability to concentrate heat in particular tissues, is their similar adaptation to high speed swimming in pelagic, nutritively dilute, waters where food is patchily distributed (Stevens and Neill 1978). It remains a paradox that so much of the broad architecture of fish circulatory systems remains little altered through phylogenetically widely divergent groups whilst tunas, swordfish and mackerel sharks have so radically rerouted theirs as part of their parallel evolution of regional warm-bloodedness.

8

VENOUS RETURN AND VENOUS PUMPS

Venous pressure

Venous pressures in fish are low and are often below or close to ambient (Table 8). Only in the hepatic portal system do they reach 10 mm Hg (Stevens and Randall 1967).

In mammals at rest, the kinetic energy of the blood in the veins is commonly a small proportion of the pressure energy. Even in the venae cavae and atria it is only 12%; in fish the very low, often subambient, venous pressures must imply that kinetic energy represents a larger proportion, and pressure energy a smaller proportion of the total. The return of blood to the heart is possible only because of the virtual absence of gravitational effects (Chapter 1) and because the ductus Cuvieri and the long veins, such as the caudal, cutaneous and lateral abdominal, have incollapsible walls and cardiac suction can lower the pressure at their central ends. Additionally, fish have a remarkable collection of auxiliary venous pumps which augment the pressure of outflows from, and steepen the gradient of, pressure across capillary beds remote from the heart. Two of these, the caudal heart and the portal heart of the myxinoids, are true hearts in the sense that their muscles serve only to power the pump. The others, such as the haemal arch pump, the branchial pump, and the so-called 'cardinal heart' of myxinoids, are propulsors, for they are powered incidentally by muscles which serve some other function such as locomotion or ventilation.

The haemal arch pump

We noted, in Chapter 3, that the vertebrae of the post abdominal spine bear ventral processes which fuse together to form the haemal arch and enclose the haemal canal (Figure 35*A*, *B*), within which lies the continu-

Table 8. *Venous pressures in fish (mm Hg)*

	Hepatic portal V.	Post cardinal V.	Hepatic V.	Caudal V.	Author(s)
Trout *Salmo gairdneri*	+ 6.9– + 9.9	—	—		Stevens & Randall 1967
Trout *Salmo gairdneri*		+ 1.4 ± 0.3			Kiceniuk & Jones 1977
Eel *Anguilla anguilla*	—	—	– 4 – + 5	—	Mott 1951
Starry flounder *Platichthys stellatus*	—	—	—	+ 2.2	Wood *et al.* 1979
Winter flounder *Pseudopleuronectes americanus*	—	—	—	+ 3.0	Cech *et al.* 1976, 1977
Spiny dogfish *Squalus acanthias*	+ 1.7– + 1.8	– 0.7– + 0.1	—	+ 1.7– + 1.9	Satchell 1971
Carpet shark *Cephaloscyllium isabella*	—	—	—	+ 1.6	Satchell & Weber 1987

ation of the median dorsal aorta, caudal vein and one or two secondary veins. Just caudal to the rostral margin of each vertebra, a segmental artery arises and passes outwards; the artery supplies many capillaries to the peripheral fillet of red muscle, and fewer to the white muscle (Chapter 6, Figure 27*B*, *C*). Blood from these is collected into segmental veins, which enter the caudal vein. Their openings are guarded by a prominent pair of ostial valves (Figure 15*B*, *C*, Figure 35*B*). In elasmobranch fish such as the Port Jackson shark, the spiny dogfish and the white pointer shark, valves occur in the segmental arteries also. They are thicker than those in the veins and are directed away from the opening of the vessel, towards the capillary beds (Satchell 1965).

In restrained elasmobranchs, lateral movements of the post abdominal trunk cause abrupt elevations of caudal vein pressure. Moreover, deflections away from the midline, whether to left or right, both cause similar elevations. These observations are consistent with the view that contraction of the myotomal muscle compresses the capillaries and veins and force blood past the ostial values into the caudal vein. Additionally, in those species which also possess arterial valves, these close and prevent blood

Figure 35. The haemal arch pump. *A*, Transverse section of post abdominal trunk of teleost fish showing main blood vessels. *B*, Details of haemal arch area. C.V. caudal vein, D.A. dorsal aorta, D.C.V. dorsal cutaneous vein, H.A. haemal arch, H.C. haemal canal, L.C.V. lateral cutaneous vein, Sc.A. secondary artery, Sg.A. segmental artery, Sg.V. segmental vein, Sv.V. subvertebral vein, i.e. a secondary vein, V. valves, V.C.V. ventral cutaneous vein, Vt. vertebra.

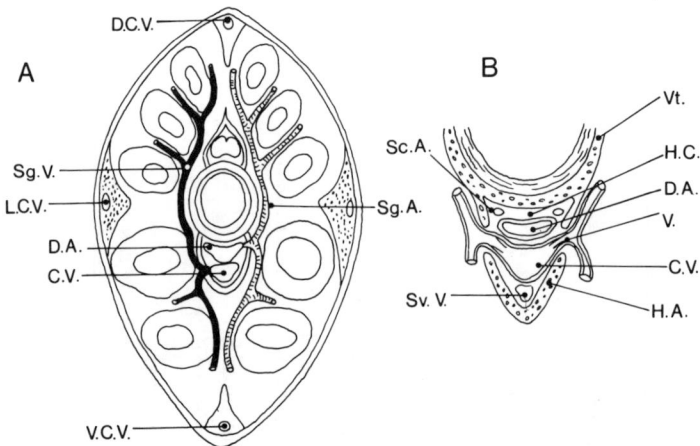

from re-entering the aorta. This interpretation is supported by experiments in which the whole post abdominal trunk of a Port Jackson shark was perfused through the dorsal aorta, and the perfusate collected and weighed (Figure 36), whilst swimming movements were evoked by electrically stimulating the spinal cord. During one minute's stimulation, flow through the preparation was increased by 46%. Moreover, the outflow showed discrete increases as the muscle vessels, to left and right, were compressed and spurts of perfusate were forced from the caudal vein. The haemal arch pump is a mechanism which exploits the muscle contractions that generate the backwardly moving waves of flexion which propel the fish forward. It is presumably a very ancient mechanism; the haemal arch is an intrinsic part of it and these skeletal elements are recognizable in the fossilized skeletons of such fish as *Xenacanthus*, a freshwater shark from the Carboniferous of 300 million years ago. Hagfish, such as *Myxine*, lack a spine and haemal arches and the caudal vein is very short and confined to only the most posterior part of the trunk. Flood (1979) has described the vascular supply to the parietal muscles of *Myxine* and it seems likely, from his illustrations, that intramuscular veins would be compressed and

Figure 36. Movement and outflow from the caudal vein in the isolated perfused trunk of the Port Jackson shark. *A*, At rest, *B*, during 1 min of electrically evoked movement, *C*, in the period following B. Redrawn after Satchell 1965.

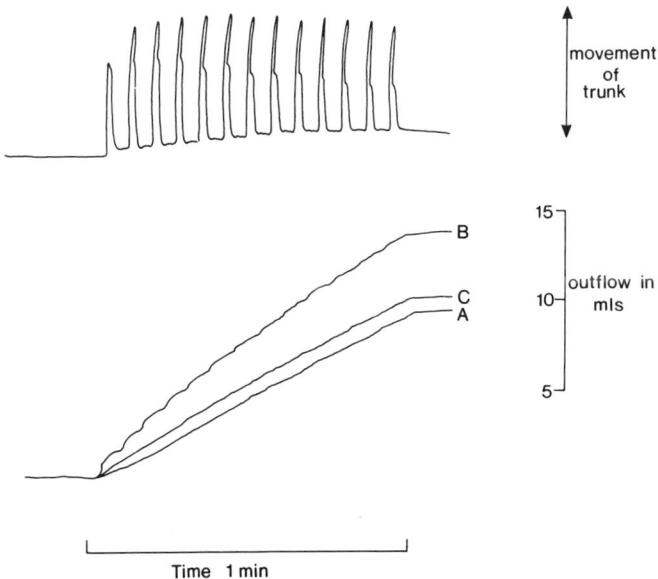

emptied into the posterior cardinal veins, but the mechanism is clearly less perfected than in the true fish.

The branchial pump

The branchial pump consists of the anterior cardinal sinus, which, we have noted, has a prominent ostial valve guarding its opening into the ductus Cuvieri, and the branchial veins, which extend up the gill bars and lie just beneath the surface of the pharyngeal epithelium. Valves occur at the openings of the branchial veins into the floor of the anterior cardinal sinus. The thin vessel walls are easily compressed during the expulsive phase of each ventilation, and blood can be seen to issue past the valves into the sinus when this is opened in a live fish. The compression of the pharynx during the expulsive phase also diminishes the size of the sinus and forces blood backwards past its ostial valve, into the ductus Cuvieri. This is a propulsor that harnesses some of the muscle power of the ventilatory pump. Cardiorespiratory synchrony (Chapter 11) may augment the efficiency of this pump, by synchronizing the fall in central venous pressure due to ventricular ejection, with the rise in branchial vein pressure due to the expulsion of water from the pharynx.

The 'cardinal heart' of the myxinoids

The term 'cardinal heart' is so firmly in the literature (Cole 1925, Brodal and Fänge 1963) that it is probably wise to retain it. An examination of the structure in a living hagfish shows clearly, however, that it is a propulsor. Water is propelled into the pharynx of a hagfish through the agency of a velum, which acts as a ventilatory pump. The velar membrane arises from the dorsal midline of the velar chamber; it is free at its posterior end, and is wafted up and down like a matador's cloak. Its movement is complex and is controlled by four muscles and a series of cartilages (Johansen and Strahan 1963). Above the velar chamber lies the hyoid plate and between this and the olfactory organ is an extensive blood sinus, the hypophysio–velar sinus (Figure 37A, B). Two muscles which insert onto the plate, the veloquadratus and the velospinalis (Figure 37B), lie within the sinus (Cole 1906–7). They impart movements to the velum but in so doing compress the sinus by moving the plate up and down against the olfactory organ. The sinus communicates via a small vessel, the subcutaneous anastomosis, with the rostral end of the subcutaneous sinus, a large blood space which encircles the animal beneath the skin (Chapter 12). Valves in the openings of these communicating vessels ensure that flow is from the subcutaneous sinus, into the hypophysio–

velar sinus. Blood leaves it through valved vessels which open into a separate part of the anterior cardinal vein of each side; this is separated from the main part by a pair of prominent valves and is the structure to which the misnomer 'cardinal heart' was originally applied. The veloquadratus is in three fasicles; the largest of these is an antagonist of the two smaller portions. The rhythmic contractions of these muscles contribute to the complex movement of the velum, and in the process of doing so, compress and expand the velar sinus. The cardinal propulsor, in that it returns blood from the subcutaneous sinus to the primary veins, performs the same role at the rostral end of a hagfish, as the caudal heart does in the tail. Cole (1925) notes the similarity of its muscles to those of the caudal heart for in both the fibres are deep red and vascular, parallel and of small diameter.

The caudal pump

The dorsal, lateral and ventral cutaneous veins of both elasmobranch and teleost fish may extend onto the tail fin; they unite in various combin-

Figure 37. The cardinal heart of myxinoids. *A*, The veins and sinuses. *B*, The skeleton and muscles. H.V.S. hypophysio-velar sinus, Hy.P. hyoid plate, I.V. inlet valves, L.V.B. lateral velar bar, Nc. notochord, O.V. outlet valves, S.A.C.V. swollen portion of anterior cardinal vein, Sc.A. subcutaneous anastomosis, Sc.S. subcutaneous sinus, Vq.M. 1–3 veloquadratus muscle, divided into three portions, V.Sp.M. velospinalis muscle.

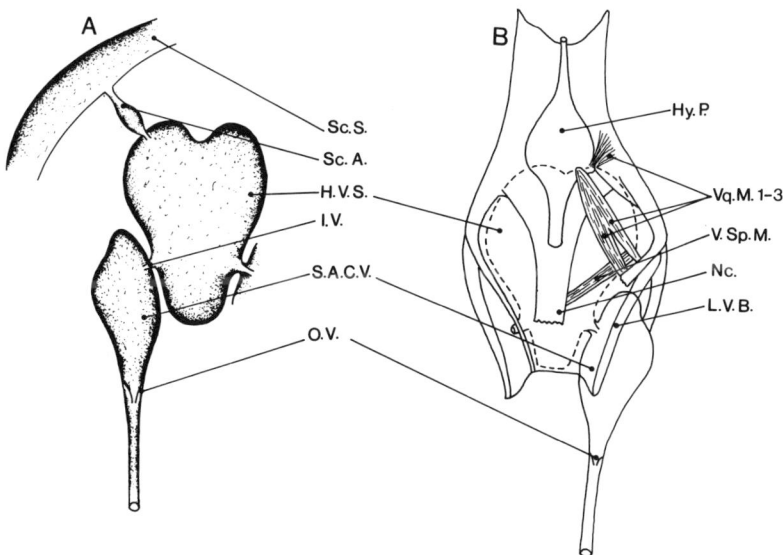

ations and supply blood to a marginal vein which encircles the tail at the junction of the membraneous and more solid central portions. This marginal vein lies superficially, just beneath the skin, and (Figure 38), in such species as the Port Jackson shark, bears valves so arranged as to direct blood towards its central region. Into it drain many small vessels from the fin lobes. Several short, valved connecting veins link the marginal vein through the superficial layer of radial muscle fibres, to a deeper vessel, the caudal sinus, which lies against the fin skeleton, beneath the caudal vein. On each side, it sends short, valved connecting vessels to enter the ventral wall of the caudal vein (Birch *et al.* 1969). The radial muscles are attached below the sinus, to the fin spines, and insert laterally onto dense connective tissue. When the muscles of one side contract, they deflect the hypochordal lobe of the fin and enlarge the lumen of the caudal sinus; the sinus of the opposite side is stretched across the fin skeleton and compressed against the radial cartilages. Hence, as the hypochordal lobe of the tail is moved from side to side, the left and right caudal sinuses are alternately filled from the cutaneous circulation, and emptied into the caudal vein; the pump thus operates only when the fish swims.

Figure 38. The caudal pump of the Port Jackson shark. C.S. caudal sinus, C.V. caudal vein, D.C.V. dorsal cutaneous vein, L.C.V. lateral cutaneous vein, M.V. marginal vein, V.C.V. ventral cutaneous vein. Redrawn after Birch *et al.* 1969.

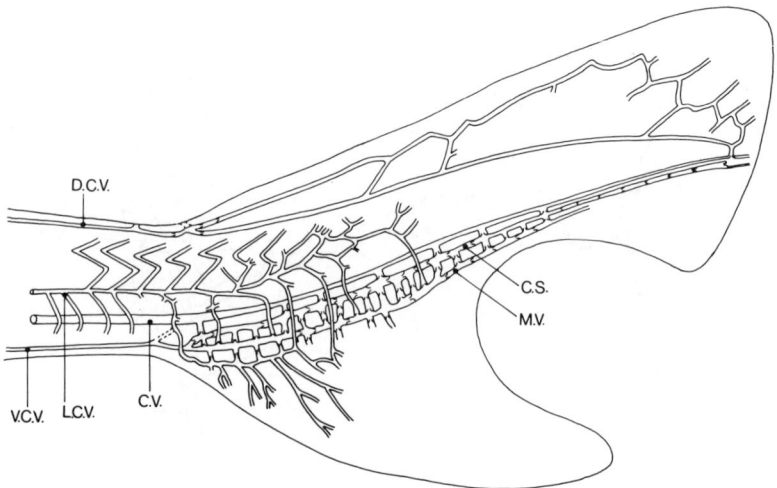

The caudal heart of the carpet shark

The blood vessels and radial muscles of the tail of the carpet shark
(*Cephaloscyllium isabella*) (Figure 39) and the smooth hound (*Mustelus
antarcticus*) are broadly similar to those of the Port Jackson shark, but
there are three significant differences (Satchell and Weber 1987). The
caudal sinus tapers from a swollen basal chamber at the base of the tail, to
a thread-like termination close to the tail margin. The valves in the sinus
all point in one direction, towards the base of the tail. There is most often
a single large vessel on each side connecting this swollen part of the sinus
to the caudal vein, although some specimens have a second connecting
vein half way along.

The carpet shark is a sedentary bottom-living species; when its tail is
examined in an unrestrained resting fish, a muscular ripple can be seen to
move rhythmically from the tip of the tail towards the base, over the line
of the caudal sinus. This appearance is caused by a serial contraction of
the radial muscles of the two sides, which pull the skin down onto the fin
skeleton, and generate a rolling wave of compression which forces blood
along the caudal sinus, past its many valves, into the swollen rostral part
and finally into the caudal vein. As the muscles of the two sides contract
synchronously they do not deflect the tail lobe, as they do when the fish is
swimming. Figure 40 shows pressures recorded in the caudal sinus, where

Figure 39. The caudal heart of the carpet shark. C.A. caudal artery,
C.S. caudal sinus, C.V. caudal vein, Co.V. connecting vein between
caudal sinus and caudal vein, D.C.V. dorsal cutaneous vein, L.C.V.
lateral cutaneous vein, M.V. marginal vein, R.M. cut-off stumps of
radial muscles; only five are shown, V.C.V. ventral cutaneous vein.
Redrawn after Satchell and Weber 1987.

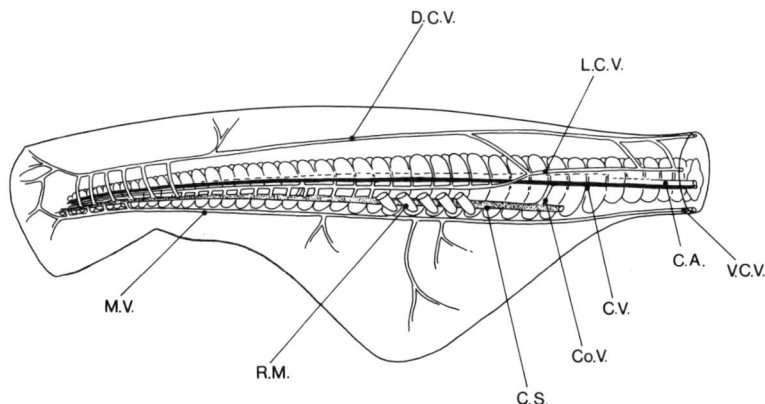

pressure is generated, in the lateral cutaneous vein, one of the veins that feeds into the pump chamber, and the caudal vein, into which the blood is pumped. The pulses of pressure in the caudal vein coincide with the expulsion of blood from the caudal sinus. The diastolic level of pressure in this is subambient and, by increasing the gradient of pressure across the secondary circulation, must enhance flow through this.

Because the caudal heart is actuated by skeletal muscle, it is paralysed by d tubocurarine; Figure 41 shows that, as this occurs, the pressure in the caudal sinus rises steadily towards that in the caudal vein. The gradient of pressure across the secondary circulation, and thus flow through it, will diminish as a result of paralysing the pump muscles.

When the fish is at rest, the radial muscles serve only to power the pump; it ought, at that time, to be regarded as a caudal heart. When the fish is swimming these muscles act as radial muscles that deflect the hypochordal lobe of the fin. The pump then must act like that of the Port Jackson shark, and should be regarded as a propulsor. The structure is intermediate between a propulsor and a heart, and shows that the distinc-

Figure 40. Pressures in the caudal vein, the lateral cutaneous vein and the caudal sinus of a resting unanaesthetized carpet shark. Redrawn after Satchell and Weber 1987.

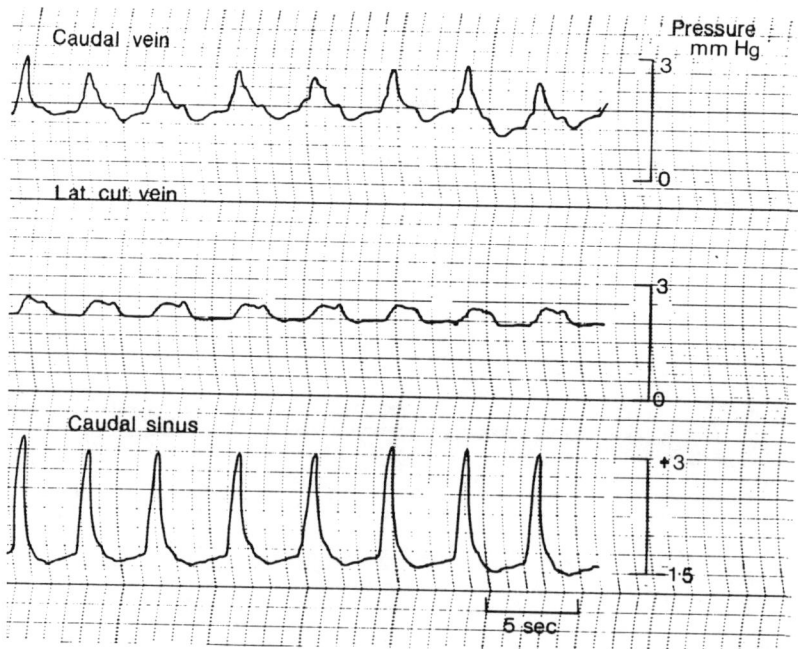

tion hitherto drawn between the two is not clear cut. The significant change evolved in the carpet shark is the ability of the motor neurones of the radial muscles, located in the end of the spinal cord, to continue to discharge when the fish is at rest, and to discharge synchronously on the two sides in a caudal–rostral sequence, rather than alternately. This ensures that the hypochordal lobe does not move and the contraction results only in the simultaneous compression of the left and right sinuses against the fin skeleton.

The caudal heart of the eel

The caudal heart of the eel (Favaro 1906) is a tiny structure, only a few millimetres long, located near the tip of the tail, below the last vertebra. It is not a paired structure (Figure 42), but consists of just two chambers, one each side of the fin skeleton, which communicate through a gap, the hypural foramen, between the hypurals, i.e. the ventral fin radials. The chamber on the right receives all of the inflows from the cutaneous veins and fin drainage; it was termed by Favaro, the 'atrium'. It has a valved

Figure 41. Similar records as in Figure 40, recorded at slower speed, following the injection of d tubocurarine. The slight pulse remaining in the caudal vein, following paralysis of the caudal heart, is that of the true heart, directly communicated within the haemal canal from the dorsal aorta. Redrawn after Satchell and Weber 1987.

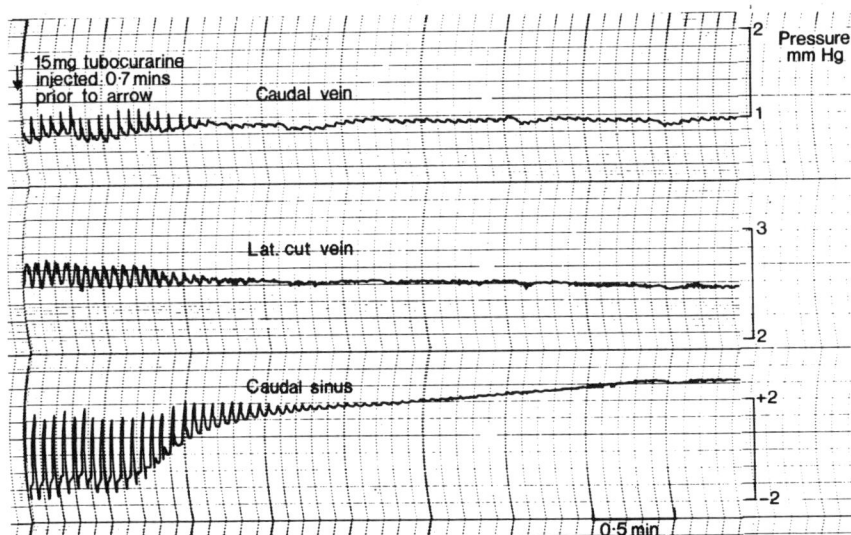

opening to the left chamber, the 'ventricle', which in turn has a valved opening into the caudal vein.

The chambers can be compressed against the fin skeleton, by delicate strands of skeletal muscle which originate on the vertebra above, and are inserted onto the hypurals. They contract alternately, so that blood in the right chamber is forced through the hypural opening into the left chamber, where, in turn, it is forced into the caudal vein. The muscle fibres are innervated from motor neurones in the last but one segment of the spinal cord, and the heart ceases to beat if their axons are cut. Mislin's (1969) study emphasizes the independence of the motor centre of the caudal heart from other motor centres in the cord. The caudal heart ceases beating at 11 °C, at which temperature ventilation continues at a slow rate. It is clear that the eel caudal heart is powered by much reduced radial muscles; they contract alternately, as they would in a swimming elasmobranch, but they are so small that no movement of the fin occurs.

Figure 42. The caudal hearts of the hagfish and the eel, in dorsal view. C.V. caudal vein, C.F.V. caudal fin vessels, Hy.F. hyural foramen, L.Ch. left chamber (ventricle), L.C.V. lateral cutaneous vein, M.V. marginal vein, R.Ch. right chamber (atrium), V.Sc.S. vein from subcutaneous sinus. Redrawn partly after Favaro 1906.

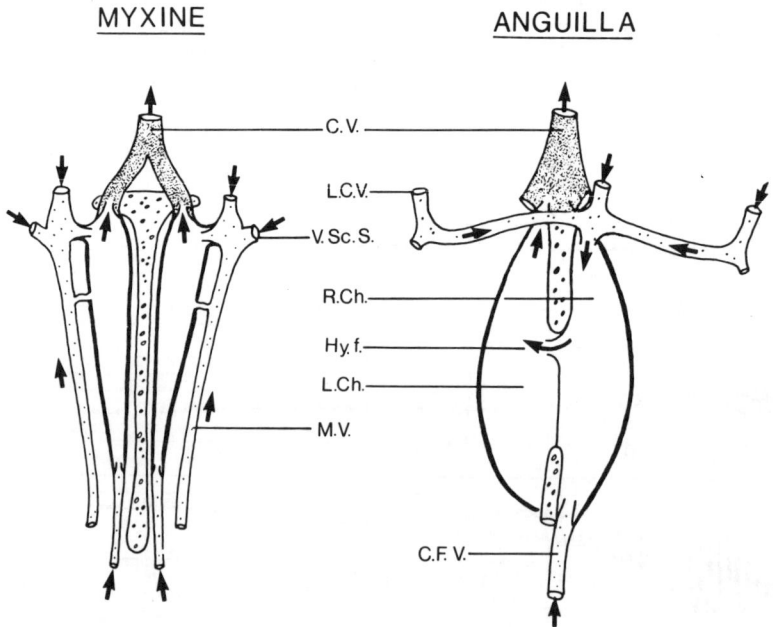

The caudal heart motor centre lies directly above the urophysis (Chapter 6). Chan (1975) reports that in the eel, urotensin 2 increases the rate of caudal heart beat, which suggests that this hormone may have a role in stimulating flow through the cutaneous circulation. Davie (1981a) has recorded bursts of action potentials, in sequence with the caudal heart beat, from afferent fibres which run from the caudal heart of the eel, back into the cord; there may thus be a nervous regulation of its rate of beat.

Caudal hearts occur in many species of teleost fish, but all of the Acanthopterygii, the spiny-finned fish which include such groups as the flounders, perch and wrasse, lack them. In the tench both left and right chambers receive inflows from the cutaneous vessels, and both communicate with the caudal vein (Hyrtl 1843). They also communicate with each other through the hypural foramen, but the opening is not valved, so the heart is double and symmetrical around the midline.

The caudal heart of the hagfish

Many features of hagfish anatomy single them out as primitive (Chapter 12) but, curiously, their caudal heart is not unlike that of teleosts. It

Figure 43. A lateral view of the caudal heart of *Myxine*. C.H. caudal heart, C.H.M. caudal heart muscle, C.V. caudal vein, D.C.V. distal caudal vein, K.M.P. knob of median fin plate, L.M.G. last of the series of mucous glands, M.N. motor nerve, M.P. median plate, M.V. marginal vein, Nch. notochord, R.V. radial vessels, Sp.C. spinal cord, V.Sc.S. vessel from subcutaneous sinus. Redrawn after Kampmeier 1969.

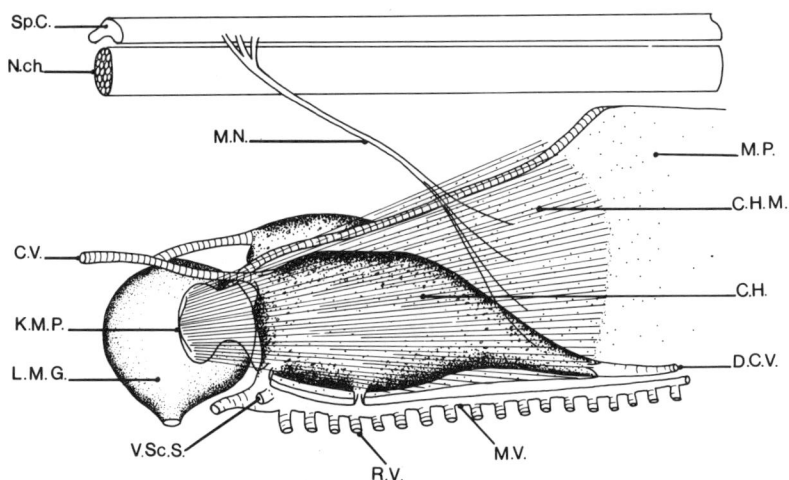

differs in that the two chambers are quite separate (Figure 42), and do not communicate across the midline. Moreover, the caudal heart muscle, which forms a thin sheet on each side, is not inserted onto the vertebrae, but onto a separated part of the median fin skeleton (Figure 43), which is pulled from side to side like a swing-door (Satchell 1984a). Jensen (1965) has recorded short trains of 4–8 muscle action potentials, characteristic of skeletal muscle activation, from the surface of this muscle, in the Pacific hagfish.

When the left muscle contracts, the left caudal heart chamber has space to expand, in the curved enclave between the muscle and the bent cartilage. The right chamber is pulled across the opposite curve, and compressed. As the opposite muscle contracts, the emptied right chamber fills, and the left chamber, now full, is compressed and emptied. Thus each whole cycle of left–right movement should cause two pulses of pressure in the caudal vein, rather than the single pulse seen in the carpet shark, in which the two muscles contract synchronously. Two such pulses can be recorded in the caudal vein (Figure 44), which suggests that this speculation of how it functions is correct. The fact that the rostral end of the cartilage between the chambers is free to move ensures that the tail remains still despite the fact that the muscles contract alternately.

The hagfish caudal heart is not continuously active (Figure 45) but starts to beat half a minute or so after a period of active swimming; hagfish in captivity exhibit periods of swimming, followed by periods of rest. After the initial burst of caudal heart activity, there are successive

Figure 44. Pressure pulses in the caudal vein of *Myxine* caused by the caudal heart. Top = pressure trace, bottom = mechanogram from surface of caudal heart. Redrawn after Satchell 1984a.

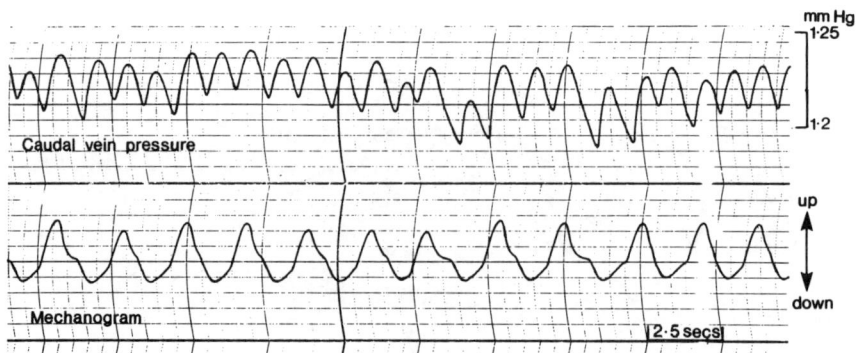

short periods of beating, which gradually die away, until the next period of swimming occurs. Vogel (1985b) emphasizes that the caudal heart is not a lymph heart, as it was assumed to be by earlier anatomists, but is quite specifically the auxiliary heart of the secondary blood system. Its position in the tail is a strategic one; it is located at a site, far away form the *vis a fronte* of the heart, at the junction of the secondary veins with the primary venous system. We noted, when discussing the secondary blood system, in Chapter 6, Davie's (1982) estimation of the rate of pumping of the caudal heart of the eel. We will return again, in Chapter 12, to the topic of whether the subcutaneous sinus of the hagfish should be considered as homologous with the lateral cutaneous vein of the true fish, and thus part of a secondary blood system.

The portal heart of the hagfish

Hagfish are unique in the possession of an auxiliary heart in the course of the hepatic portal vein, which elevates the pressure of the blood from the intestine prior to its course through the sinusoids of the liver (Figure 46A). Moreover, the heart is powered by cardiac muscle in every way like, in its ultrastructure, that of the true heart nearby. The portal heart is like a small atrium, and generates its own electrocardiogram, with a P and a T wave (Figure 46B). Its contraction creates a pressure of 2–3 mm Hg,

Figure 45. Record of rate of caudal heart beat of *Myxine* following a period of swimming. Solid circles, 1 °C, hollow circles 14 °C. The arrow indicates a second short period of swimming (5 min) which occurred in the course of the 14 °C record. Redrawn after Satchell 1984a.

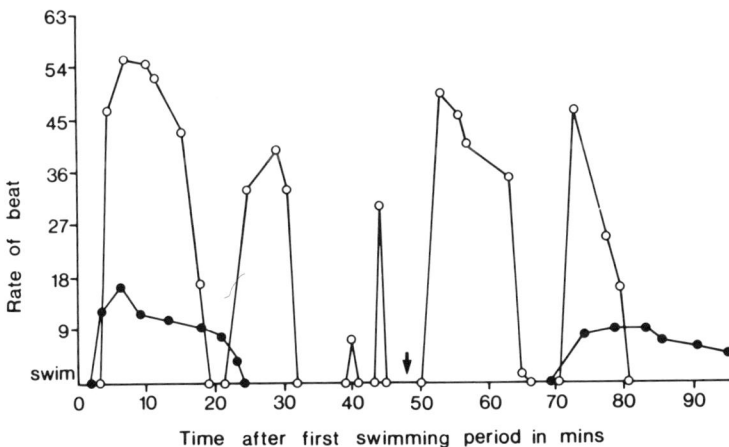

Figure 46. The portal heart of *Eptatretus*. *A*, The heart in longitudinal section. *B*, Simultaneous records of (top) ECG and, (bottom)
intracardiac pressure. Calibration, pressure in mm Hg. Br.H.P.V.
branches of hepatic portal vein, C.M. cardiac muscle, H.V. hepatic
vein, S.I. sinus intestinalis, P, T, the P and T waves of the portal heart
electrocardiogram. *B*, Redrawn from Davie *et al.* 1987.

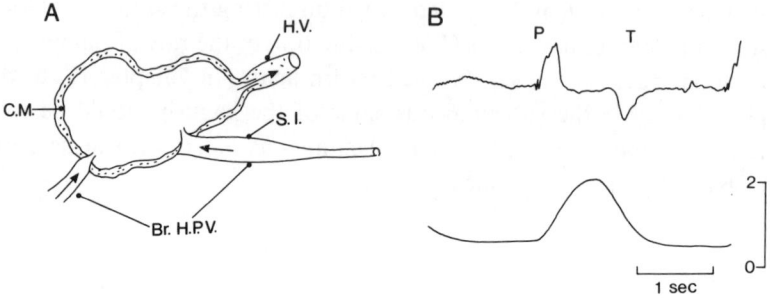

much as does the atrium of the true heart (Figure 6 At.). In both the
Atlantic hagfish and the New Zealand hagfish, the region of the portal
vein behind the portal heart, termed the sinus intestinalis, contracts in a
wave-like manner, and seems to have the role of a sinus venosus. No
element of the portal heart ECG has, with certainty, been associated with
this. The portal heart and branchial heart beats are not coordinated,
which is to be expected as they receive no innervation.

Auxiliary venous pumps are probably advantageous to fish for they
enable the unstressed venous blood volume to be mobilized. In mammals
this can be achieved by venoconstriction, but the somatic veins of fish lack
autonomic nerves. The secondary microvessels of the scales have delicate
walls, and the scale epithelium must impose little mechanical restraint.
Thus a volume of blood is contained within them at a pressure which must
be very close to ambient. We have already noted that the secondary
circulation may contain more than twice as much of the plasma as the
primary vessels, and caudal hearts are possibly of importance in regulating the flow of blood from the secondary into the primary veins.

9

THE AUTONOMIC NERVOUS SYSTEM

Introduction

A trout, during exercise, may increase its cardiac output threefold and augment the flow of blood to its red muscle from 1.14 to 16.21 ml min^{-1} kg^{-1} (Randall and Daxboeck 1982). In Chapter 2 we saw that a greater return of blood to the heart will increase its output, through the Starling law, and in Chapter 3, hyperaemia, which increases the flow of blood to active tissues, was described. The greater pressure in the efferent branchial arteries that occurs during exercise may in turn increase the surface area perfused with blood of individual gill lamellae; it may also recruit additional lamellae into the circulation. Both of these responses can be exspected to increase the ability of the gills to take up oxygen.

All four mechanisms are intrinsic to the tissues, cardiac muscle, or vascular smooth muscle, which respond in these ways, and they help to maintain an adequate circulation under a variety of external conditions. Such mechanisms were, presumably, sufficient on their own in ancestral vertebrates. In hagfish they must be particularly important because there is no autonomic innervation of the heart; blood vessels too are minimally innervated if at all and there are no innervated central stores of catecholamines.

In more active fish the demands of different organs and tissues may, however, be in conflict. The heart may be unable to supply sufficient blood to the skeletal muscles unless the flow to the gut and other viscera is temporarily curtailed. Under conditions of severe hypoxia the skeletal muscles may prove to be better able to endure a spell of glycolysis than the brain or the retina. But glycolysis yields less energy per molecule of substrate than aerobic oxidation and it may be necessary, temporarily, to increase the level of circulating blood glucose if energy output is to be maintained. There arises a need to integrate the flows to different tissues with the ability of the heart to supply them. This overall control of the circulation is brought about by the autonomic nervous system and, in the advance from the hagfish to the elasmobranch and teleost grades of complexity, we can see its progressive evolution as more and more organ systems are captured by it. Those wishing to extend their understanding of this topic beyond the scope covered in this Chapter, should read Nilsson's superb monograph on the subject (Nilsson 1983).

The parasympathetic and sympatho-adrenal systems

The autonomic nervous system comprises all of the outputs of the central nervous system other than those to skeletal muscle. It differs from the somatic nervous system in several features of anatomy and function.

The pathway between a spinal motor neurone and its skeletal muscle fibres is bridged by the axon of a single motor neurone. In the autonomic nervous system the equivalent pathway is bridged by two nerve cells. The first one, with its cell body in the brain stem or spinal cord, is termed the preganglionic neurone and it synapses with the second, the postganglionic neurone, in an autonomic ganglion which lies outside the central nervous system. The preganglionic axons are myelinated, and thus appear white. From the ganglion, unmyelinated, grey, postganglionic axons run to the effectors which may be pace-maker cells, cardiac muscle cells, smooth muscle fibres or secretory cells.

The separate skeletal muscle fibres of higher vertebrates are almost always innervated by single motor nerve endings. In the autonomic nervous system many, but not all, viscera are innervated from the two subdivisions of the autonomic nervous system, termed the parasympathetic and the sympathetic. In fish, all of which lack any sacral parasympathetic nerves, the most significant parasympathetic outflow is the vagus nerve. This arises via multiple roots from preganglionic cells in the medulla and passes backwards to innervate the gills, heart and many other viscera far back into the abdomen (Figure 47).

In teleost fish the sympathetic preganglionic fibres exit from their many cell bodies in the spinal cord via the ventral motor nerve roots, as white rami, and almost immediately synapse with their postganglionic fibres in ganglia which are linked longitudinally to form a chain. The two chains, one each side of the spinal column, are linked across by ladder-like connections. From the chain ganglia, unmyelinated fibres run to the various structures they innervate; these include the heart and the smooth

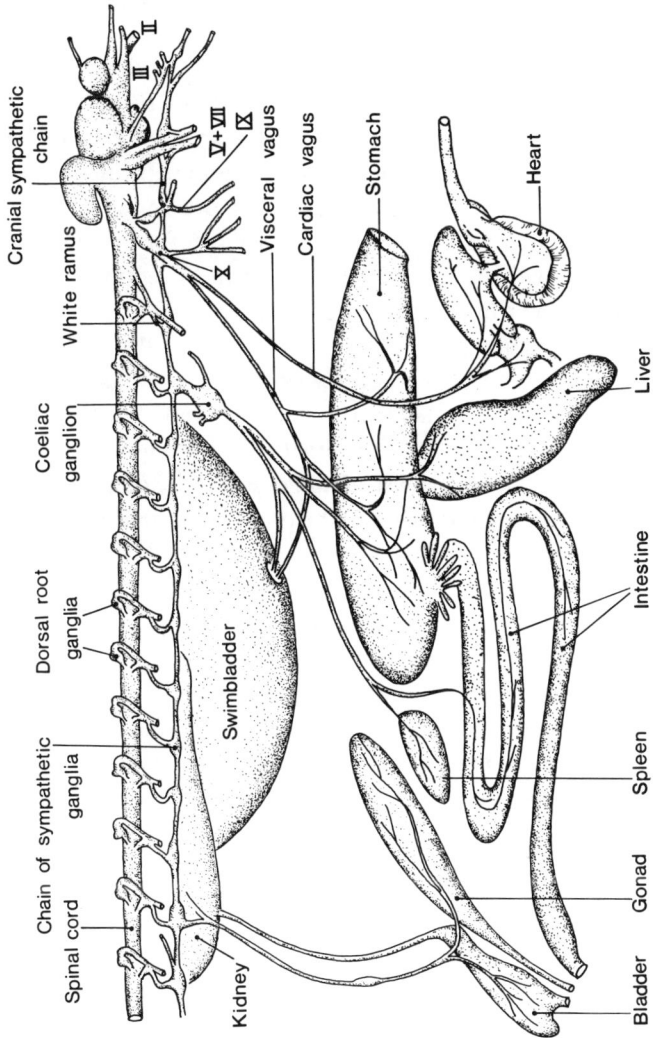

Figure 47. The autonomic nervous system of a teleost fish. The cranial nerves are designated with Roman numerals. Redrawn after Nilsson 1983.

muscle of the blood vessels of the gut and liver. The sympathetic chains extend into the head and carry postganglionic fibres destined for the gill vessels. In some fish sympathetic adrenergic fibres enter the vagus and are distributed to the viscera along its many branches. The lateral chains of elasmobranchs are much less regularly developed, and do not extend into the head nor, it seems, do sympathetic postganglionic fibres run directly to the heart.

There is thus a broad distinction in gross anatomy between the two systems. In the parasympathetic, almost all of the distance between the preganglionic cell body and the effector is spanned by myelinated preganglionic fibres; the postganglionic cell bodies are located in the tissue they innervate, and their axons are thus very short. In the sympathetic system, the preganglionic myelinated fibres are short, for they have only to reach to the chain ganglia close to the exit of their ventral roots. Most of the distance to their final termination is traversed by the grey unmyelinated postganglionic fibres.

We are on less certain ground when we consider our third distinction. It is broadly true that the two systems operate by different mediators. Whereas the postganglionic terminals of the synapse between the pre- and the postganglionic neurones, all liberate acetylcholine, as also do those of the postganglionic terminals of the parasympathetic system, the postganglionic terminals of the sympathetic system liberate catecholamines, either noradrenaline or the methylated form of this, adrenaline. Moreover, whereas liberated acetylcholine is quickly removed from the synaptic area by the enzyme acetylcholinesterase, catecholamines are removed by reuptake into the nerve terminals which liberated them. However, it has been increasingly realized that cholinergic neurones occur also in the sympathetic system, and, it seems, some neurones can secrete both catecholamines and acetylcholine. We will look at the evidence for this when we consider the control of the spleen.

This broad division of autonomic neurones into two categories, adrenergic and cholinergic, has to some extent been obscured by the discovery of a variety of other neurones secreting other mediators. Various neuropeptides have been identified; in the little skate (*Raja erinacea*) and starry ray (*R. radiata*), neuropeptide Y, a 36 amino-acid peptide, occurs in fibres innervating the ventricle, conus and systemic arteries (Bjenning *et al.* 1989). The functional roles of these and other peptide-containing neurones have, as yet, been little investigated in fish.

The sympathetic system has an additional feature. The two catecholamines are liberated, not only from the postganglionic terminals at sites very close to where they will exert their effects, but also directly into the blood stream from masses of tissue specialized for their synthesis. This, the so called 'chromaffin tissue' is the equivalent of the adrenal medullary tissue of higher vertebrates; in most fish it is not associated with the cortical tissue to form a compound gland. In many teleosts it occurs as masses in the walls of the posterior cardinal veins where these pass through the pronephros or head kidney, a structure we noted in Chapter 5. The chromaffin cells are innervated by preganglionic fibres and have been regarded as representing postganglionic cells in which the fibre has been repressed. In elasmobranchs, a mass of chromaffin tissue and post-ganglionic neurones lies each side in the dilated part of the posterior cardinal sinus; from it postganglionic fibres run to the visceral blood vessels. These paired structures, termed the axillary bodies, are well placed to pour out catecholamines directly into the sinus blood. The *vis a fronte* of the heart will ensure that they are aspirated directly into the heart, where they can mediate adrenergic effects on the myocardium, and quickly pass onwards to the vessels of the gills.

  In the cod it is clear that the catecholamines which appear in the blood in response to hypoxia are almost all derived from the chromaffin tissue in the pronephros. The topic will be discussed in greater detail in Chapter 11 but we can here note that electrical stimulation of spinal nerve roots 1–4 in the cod causes a rise in blood catecholamine levels and cutting these nerve roots abolishes almost all the response to their stimulation (Wahlqvist and Nilsson 1980). Nilsson *et al.* (1976) report that when the chromaffin tissue of cod was loaded with labelled adrenaline, a dose-dependent release of label occurred when acetylcholine was injected. This mediator would excite the ganglionic synapses between the autonomic preganglionic fibres and the chromaffin cells. Hexamethonium, a ganglion-blocking drug, reduced this release. There is thus sound experimental evidence for the presumed role of this pathway.

Adrenergic and cholinergic receptors

This coupling of the chromaffin tissue into the sympathetic system enables the effects of its mediators, the catecholamines, to be more widely dispersed. A sympathetic response, such as the contraction of blood vessels or an increase of cardiac output, can be evoked rapidly by the direct localized liberation of the catecholamine from nerve terminals onto the effectors, and more slowly, over a longer period, by the outpouring of catecholamines into the blood.

  Catecholamines, liberated into the circulation by activation of the chromaffin tissue, find their way to all the effectors that are sensitive to them, and such a system seems better suited to bring about some widely distributed response than to mediating some localized change. Some discrimination is, however, possible, because the membranes of effectors such as pace-maker cells and smooth muscle fibres, have particular receptor proteins, which interact specifically with mediators such as acetylcholine or the catecholamines.

A number of these receptors are now known; two of particular importance in the autonomic system of fish provide for alternative responses to catecholamines and are termed α and β adrenergic receptors. They are distinct proteins and are the products of separate genes. They can each be specifically excited, or blocked, by particular drugs and these have been widely used by investigators as analytic tools. Alpha mediated responses are, for example, activated by noradrenaline and blocked by phentolamine and yohimbine. Beta mediated responses are activated by isoprenaline and blocked by propranolol. Some agonists such as adrenaline and noradrenaline have both α and β effects. If, as often happens, the α effect predominates, adrenaline may constrict a particular vascular bed, but the blocking of the α receptors with phentolamine may reveal an underlying β effect which then brings about vasodilation. In most vascular smooth muscle, α activation causes contraction and constriction of the vessel; β activation causes relaxation.

Beta receptors can in turn be subdivided into two types; in general it is true that β_1 receptors mediate the inotropic and chronotropic responses of the heart and β_2 the vasodilatory responses of vascular smooth muscle. β_1 and β_2 receptors can also be distinguished by their sensitivity to different agonists and antagonists. Wood (1975) reports that the β receptors of trout branchial vessels show sensitivities to isoprenaline (ISO), noradrenaline (NAd) and adrenaline (AD), of ISO 100: NAd 3.3: AD 1, a feature in which they resemble the β_1 receptors of the mammalian heart. A. P. Farrell (personal communication) reports the presence of β_1 receptors in coronary artery rings of the long-nosed skate (*Raja nasuta*) and their response to isoprenaline is reduced by atenolol, a β_1 antagonist.

Catecholamines can increase the stroke volume and rate of isolated perfused hearts. The compact muscle of the carp ventricle shows sensitivities to ISO, AD and NAd (Temma *et al.* 1986) which indicate that its increase in rate in response to these catecholamines is mediated via β_2 receptors. In mammals it is known that cardiac responses can be mediated by both β_1 and β_2 receptors.

The liberation of catecholamines from the postganglionic nerve terminals is blocked by the drug, bretylium. It has been used as a means of distinguishing between neural and humoral control (Smith *et al.* 1985). These authors compared the blood pressure reduction in cod caused by bretylium, which might be expected to block adrenergic nerve terminals, with that caused by the α antagonist phentolamine, which would block the vasoconstrictor effect of both adrenergic nerves and circulating catecholamines.

The transmission of activity across the cholinergic synapses in the chain ganglia can, in some species, be blocked by hexamethonium. The action of acetylcholine on the heart, and on the gill vessels, is blocked by atropine. Many other agonists and α and β blockers are known to pharmacologists, and the reader in pursuit of further detail should have recourse to a modern text such as that of Rang and Dale (1987). The modest purpose of these paragraphs is only to make intelligible the text that follows in Chapters 10 and 11.

Antagonism and alliance in the autonomic system

The parasympathetic and sympatho-adrenal systems act, to some extent, in opposition to each other. Four examples of autonomic regulation of the vascular system illustrate variations in the extent of this.

(1) Wood (1975) has shown, in the trout, that acetylcholine causes a dose-dependent constriction of the gill vasculature and $10 \ nmol \ l^{-1}$ increases gill resistance by 250–300%. Adrenaline and noradrenaline dilate gill vessels and $100 \ pmol \ l^{-1}$ cause a 60% reduction in vascular resistance; the responses are opposed, the one to the other.

(2) In the heart the opposed effects are not quite so clear. Acetylcholine and electrical stimulation of the vagus slow the heart. Adrenaline and noradrenaline increase the force of contraction of the isolated and the *in vivo* heart. They only rarely cause cardiac acceleration in resting fish as reflex slowing in response to increased cardiac filling occurs. They are particularly important in stress. Catecholamines are liberated following stressful exercise, and they powerfully enhance the Starling response to the increased venous return caused by skeletal muscle pumps (Farrell *et al.* 1986b). The coronary circulation of the Atlantic salmon shows opposite responses to the α and β stimulation of injected catecholamines (Farrell and Graham 1986); α constrictor effects predominate, and can be blocked by yohimbine; isoprenaline, a β agonist, causes dilation and can be blocked by propranolol. Isolated strips from the larger coronary arteries of the long-nosed skate have a mixed population of α and β_1 receptors. The unpublished study of Small (A. P. Farrell, personal

communication) reports that β_2 receptors are lacking from the coronary vessels of trout, for the specific β_2 antagonist, ICI 115856 is without effect. Fish thus differ from mammals, in which β_1 receptors predominate in the larger coronary and β_2 in the smaller resistance vessels (Zuberbuhler and Bohr 1965).

(3) The swimbladder is an organ in which α and β adrenergic, and cholinergic, mediators interact in a complex way to bring about its inflation and deflation (Chapter 6). There is a separate gas gland ganglion and cholinergic fibres innervate the gas gland mucosa. Gas secretion in the goldsinny wrasse (*Ctenolabrus rupestris*) is inhibited by cutting the vagus or administering cholinergic-blocking drugs such as atropine (Fänge *et al.* 1976). Electrical stimulation of the sympathetic nerves to the swimbladder of the cod, and α adrenergic agonists, constrict its arterioles, a response which is reduced by α blocking drugs (Nilsson 1972). The vagus of the eel carries some adrenergic fibres and constriction of the arterial vessels of the secretory bladder can be directly visualized when the vagus is stimulated, a response that can be blocked by rogitine, an α blocking drug (Stray-Pedersen 1970).

Relaxation of the sphincteric and vascular smooth muscle (Figure 29) and exposure of the resorbent mucosa in the oval, which is part of the deflation reflex, appears to be a β mediated adrenergic response. Isolated strips of eel bladder relax in response to β agonists (Nilsson and Fänge 1967). A study by Ross (1979b) of the capillaries of the oval in the saithe (*Pollachius virens*), using radio-labelled microspheres, demonstrated that these vessels dilated, and blood flow increased by 50%, when the bladder was inflated and the deflatory reflex elicited.

There appears also to be a vagal innervation of the muscle of the resorbent muscularis mucosae, and of the sphincter. It presumably becomes active during inflation. It is also possible that the adrenergic α and β mediated responses of the secretary and resorbent mucosa are activated by circulating catecholamines from the chromaffin tissue.

(4) The spleen, in fish as in higher vertebrates, is divided into a red outer cortex from the tissue of which erythrocytes and thrombocytes are generated, and a paler inner medulla which gives rise to leucocytes (Chapter 5). It is surrounded by a capsule containing smooth muscle, and this extends inwards to form trabeculae. When this muscle contracts, red cells, contained in spaces within the spleen, can be expelled via the splenic vein into the blood system and this increases the haematocrit of the circulating blood. The spleen of teleosts contracts strongly during severe exercise (Yamamoto 1988).

The spleen is innervated by a sympathetic nerve which runs from the coeliac ganglion (Figure 47) and when this is electrically stimulated splenic constriction occurs and red cells are released into the vein (Nilsson and Grove 1974). The nerve contains both cholinergic and α adrenergic postganglionic fibres and when acetylcholine or noradrenaline are perfused through the spleen, both cause the smooth muscle to contract. Alpha adrenoceptor antagonists such as phentolamine and cholinergic antagonists such as atropine can separately block these two responses.

When the coeliac ganglion is stained specifically to reveal adrenergic neurones, these turn out to be almost all the neurones in it, and this would seem to suggest that cholinergic neurones are scarce, a conclusion not supported by the vigorous response to cholinergic stimulation. Moreover, 6-hydroxydopamine, which is taken up by, and specifically destroys adrenergic terminals, blocks the response both to perfused catecholamines and to acetylcholine (Holmgren and Nilsson 1976). After such treatment stimulation of the splenic nerve is without effect. The authors conclude that individual neurones must secrete both noradrenaline and acetylcholine, an idea at variance with the earlier one neurone–one mediator concept. The spleen in teleost fish is perhaps a survival of an earlier undifferentiated stage in the evolution of autonomic control mechanisms. It may be that initially two or more mediators, secreted by the one cell, evoked much the same responses and separate neurones producing opposed reactions in a common effector was a later development. Subsequently, it may be supposed, these opposite responses of stimulation and inhibition became increasingly concentrated in separate neural pathways.

Atrial natriuretic peptide (ANP)

The specific granules in the atrial myocytes of mammals (Chapter 2) store and release a circulating hormone, ANP, in the form of one or more peptides. Increased blood volume causes a mechanical stretch of the atrial wall which brings about a centrifugal movement of the granules to the surface of the myocyte, and the release of ANP into the blood (Agnoletti *et al.* 1989). ANP relaxes renal vascular smooth muscle, elevates glomerular pressure and increases salt and water execretion.

ANP also relaxes the vascular smooth muscle of other vessels if this is, in any degree, already contracted by noradrenaline, serotonin and various other agents; thus, intravenous injection of ANP causes hypotension. ANP is discussed, at this point, as an example of a regulative vascular hormone which, in fish, interacts with the autonomic nervous system and is secreted from one particular tissue of the vascular system to act on another. ANP may well be more ancient than the autonomic nervous system, which is poorly developed in the hagfish, for ANP has been detected in the brain and heart of *Myxine*, and binding sites for it occur in the aorta and the nephrons of its kidney (Kloas *et al.* 1988).

In elasmobranchs, ANP stimulates fluid and solute secretion in the rectal gland, and increases chloride secretion by the chloride cells of the isolated opercular epithelium of the salt water minnow (Scheide and Zadunaisky 1988). When trout were injected with extracts of trout hearts or with synthetic ANP, blood pressure, contrary to expectation, increased (Duff and Olson 1986). It was at that time uncertain whether this was due to a direct effect on vascular smooth muscle, or to an excitatory interaction with the autonomic system. Later studies (Olson and Meisheri 1989) showed that ANP relaxed the vascular smooth muscle of isolated vessel rings in trout, as it does in mammals. It inhibited α mediated constriction of branchial vessels, and, to a lesser extent segments of the coeliacomesenteric artery. It did not change the β mediated increase in branchial permeability to water. It seems, at present, that the actions of ANP in fish are broadly similar to those in mammals.

Vascular receptors

The autonomic nervous system cannot regulate the circulation unless it receives information of a change in the variable to which it is to respond. If, when the fish swims into hypoxic water, the blood flow to some tissues is to be increased and that to others curtailed, receptors sensitive to the oxygen tension of the water must be present, and they must relay this information to the central nervous system. It is unfortunate that the term receptor is used to designate two very different entities. We now wish to consider not membrane proteins of particular configuration to which particular drugs can bind, but neural receptors, i.e. neurones or terminals of neurones, which have the capacity to discharge trains of action potentials when excited by certain stimuli to which they are specifically sensitive. Vascular receptors of this sort have been extensively studied in mammals and much of the early work on fish has borrowed ideas from this field.

Pressure or baroreceptors

Baroreceptors are the most completely studied of all mammalian vascular receptors. Their structure and location in the carotid sinus and the aortic arch, their discharge in response to different patterns of pressure change, and the reflex effects these evoke, are now widely known. Increased pressure causes baroreceptor discharges of greater rate and duration, and reflexly slows the heart and dilates arterioles, both of which lower the blood pressure. This simple negative-feedback loop forms the background understanding within which many investigators of fish circulatory physiology have worked.

Laurent and Dunel (1966), and Laurent (1967, 1974) reported the existence of fine nerve terminals in the walls of the afferent arteries of the pseudobranch in trout; they termed these the Type A receptors and noted that they are connected with myelinated nerve fibres of relatively large diameter. The pseudobranchial nerve is a branch of the glossopharyngeal or IX nerve, as is the carotid sinus nerve of mammals. If the pseudobranchial nerve was cut, these endings degenerated; they are therefore, primary afferent terminals. In isolated perfused pseudobranchs these fibres showed an increased discharge when the pressure of the perfusate was increased and reviewers have regarded them as baroreceptors.

It is necessary to remember that to demonstrate a response to a pressure change, or even rhythmic discharges in time with the arterial pulse, is not necessarily an indication that such receptors take part in a pressure-regulating reflex. In mammals many of the Pacinian corpuscles associated with the long bones, their muscles and their tendons are found to have an arterial pulse rhythm superimposed on their tonic discharge. They are mechanoreceptors of the musculoskeletal system; no one supposes they are the input to a baroreceptor reflex. The crucial evidence lies in the demonstration that specifically manipulating the discharge rate of a receptor changes blood pressure. It is questionable whether this has yet been achieved in fish.

In the spiny dogfish, a 11 mm Hg increase in the pressure of the perfusate in a cannulated branchial artery caused a brief, unsustained slowing of the heart (Lutz and Wyman 1932). Similar results have been recorded in the carp (Ristori 1970, Ristori and Dessaux 1970). The hearts of fish are sensitive to any kind of disturbance, and it is the experience of many that almost any kind of stimulus, such as disturbance of the water, vibration of the tank or a change in the illumination of the laboratory, will evoke such a response. It has been termed the startle reflex. Mott (1951) stimulated the central ends of cut branchial nerves in the eel, and showed that this slowed the heart but failed to present evidence that these fibres came from baroreceptors.

There are many reports that adrenaline injected into fish causes a rise in blood pressure. This causes a slowing of the heart, a response which is abolished by atropine (Laffont and Labat 1966, Randall and Stevens 1967). The response has been advanced as evidence of a baroreceptor reflex; it is evidence only of an inverse relationship between pressure and heart rate in which the cardiac vagus is the source of the inhibition. The study of Wood and Shelton (1980a) on trout is, however, more detailed. The loss of 10% of the blood volume, either during periods of spontaneous bradycardia, or of bradycardia induced by hypoxia, causes a fall in dorsal aortic blood pressure and an acceleration of the heart. The return of the lost blood slows the heart once more. The same response was evoked by different means. When, during a period of spontaneous bradycardia, blood pressure was reduced by very small doses of acetylcholine, the heart accelerated; it slowed again as the depressor effect subsided.

Such results suggest that a diminished return of blood to the heart reflexly leads to the acceleration of its beat. Haemorrhage of 10% of the blood volume, and a reduction in blood pressure, might both be expected to reduce the filling of the atrium. The work of Short *et al.* (1977) emphasizes that the two branches of the vagus to the heart in elasmobranchs, the branchial and visceral cardiac vagus, both contain sensory fibres from cardiac receptors, and that electrical stimulation of their central cut ends slows the heart. It is thus till uncertain whether the receptors discharged by the pressure increase evoked by catecholamines are baroreceptors located in the gill blood vessels, or volume receptors located elsewhere, perhaps in the atrium.

Hypoxia receptors

Hypoxia causes a reflex bradycardia, a topic we will explore in Chapter 11. Earlier studies (Satchell 1961) noted the extreme rapidity of the response and the fact that it occurs even when blood flow through the gills is prevented by clamping the aorta, and concluded that the receptors must be in the gills, and either face out into the water or lie very superficially. In mammals, oxygen receptors occur in the carotid bodies and in other sites away from the respiratory surface.

By irrigating particular gill slits with deoxygenated water, and thereby applying a very localized stimulus, Smith and Jones (1978) and Daxboeck and Holeton (1978) identified, in trout, a zone of receptors on the dorsal surface of the anterior face of the first gill arch. Directing the deoxygenated water at this areas, which is innervated from the IX and X cranial nerves, caused the heart to slow. In trout, too, ligating the gill artery did not slow down the rapidity of the response, suggesting that the receptors are close to or at the surface (Randall and Smith 1967). Butler and Taylor (1971) and Butler *et al.* (1977) have shown, in the dogfish, that oxygen receptors mediating reflex bradycardia must also occur in areas supplied by the trigeminal (V) and facial (VII) nerves. In coho salmon, too, section of the glossopharyngeal and vagal branches to the pseudobranch and first gill arch does not completely eliminate the response, although it greatly reduces it (Smith and Davie 1984).

Laurent (1974) identified oxygen receptors in the pseudobranch of trout; these terminals join to fine myelinated or unmyelinated axons in the glossopharyngeal nerve. Irrigating the perfused pseudobranch with a solution with a P_{O_2} of 9 mm Hg caused these to discharge and they remain the only oxygen receptors in fish to have been so investigated. Nevertheless, they are not the source of the bradycardiac reflex. Randall and Jones (1973) have shown that cutting the afferent nerve from the pseudobranch in no way diminishes the bradycardiac response.

In trout Randall and Smith (1967) note that the gill oxygen receptors, whilst the source of the bradycardiac reflex, are not those involved in the increase in ventilation which hypoxia causes. This is a separate reflex deriving from receptors elsewhere, perhaps in the medulla of the brain. When trout are immersed in water saturated with carbon monoxide the ability of their haemoglobin to transport oxygen to the tissues is blocked and ventilation increases. Such fish do not, however, show bradycardia, yet will still do so if ventilated with hypoxic water (Holeton 1977).

Gill nociceptors

There exists in the lungs of mammals a group of receptors which lie in the interstitial tissue between the alveolar walls and the pulmonary capillaries; they have been termed the juxta pulmonary capillary receptors, abbreviated to J receptors. The studies of Paintal (1955, 1957, 1969) built up a body of understanding that these receptors were a particular kind of mechanoreceptor, so positioned that incipient oedema, as might result from a Starling imbalance between pulmonary capillary pressure and colloid osmotic pressure, would discharge them by hydrating the collagen fibres around them. This, it was suggested, would cause the fibres to swell and discharge the receptors which were thus viewed as oedema receptors. They also discharge in response to the synthetic drug phenyldiguanide (PDG), when it is injected into the pulmonary circulation. PDG discharges receptors sensitive to 5-hydroxytryptamine; 5HT, a catecholamine, is sometimes a mediator at sensory synapses. In the intact animal they are also discharged by a number of inhaled irritants such as chlorine gas, phosgene, trichlorethylene, ammonia and halothane. Alloxan, in appropriate dosage, causes pulmonary oedema and evokes a slow discharge of long duration.

J receptor discharge in cats reflexly evokes a specific set of responses termed the J reflex. It comprises bradycardia, hypotension, respiratory inhibition or shallow breathing, and an inhibition of spinal motor reflexes, so that an animal injected with PDG ceases all movement for a few minutes. Paintal (1970) suggested that this was a protective reflex which guards the lung against the danger that excessive activity of the venous muscle pumps in exercise might overload the pulmonary circulation, thereby causing pulmonary oedema. Subsequent studies have revised this view and J receptors are now regarded as one of a group of pulmonary nociceptors, i.e. a receptor that discharges to a variety of stimuli which have in common that they are harmful to pulmonary tissues. Such J receptors are known to be present in a number of mammals, such as cat, rabbit, dog and man.

Satchell (1978) showed that when 200 μg kg^{-1} of PDG is injected into a spiny dogfish, a dose the same as that used in mammals, it causes bradycardia, hypotension, shallow respiration and a brief cessation of swimming movements (Figure 48). Fish, too, appear to show a J reflex in

Figure 48. The J-type reflex in the spiny dogfish. Note that following the injection of phenyldiguanide (arrow) into the branchial circulation, respiration is inhibited, the blood pressure falls, the heart slows and swimming is inhibited. Redrawn after Satchell 1976.

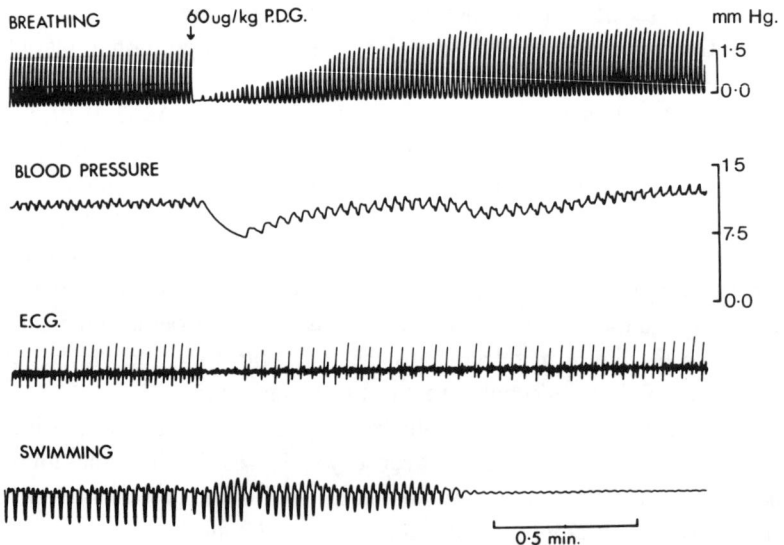

response to PDG. A subsequent study by Poole and Satchell (1979), on the perfused isolated dogfish gill, identified receptors which discharge in response to PDG, and to 5HT whether it is perfused through the blood vessels or applied in localized patches to the surface of the gill. Here too the receptor endings are accessible both from the blood vessels and the respiratory epithelium. It was possible to map the receptive fields of receptors using topically-applied PDG; some were small and extended only over a portion of one gill filament; others spread across several filaments. At one site within the receptive field, chemical stimulation would evoke a prolonged discharge with a short latency: surrounding this was a region where less intense discharges of longer latency resulted (Figure 49). The receptors also responded to gentle mechanical stimulation. The injection of 10 mg of alloxan into the gill circulation caused a visible oedema of the lamellae and increased the resistance of the gill vessels. It also caused a long-sustained and slow discharge of the receptors.

The parallels between these gill nociceptors and the J receptors of mammals are evident. In the dogfish it seems likely that they are discharged by toxicants in the water (Satchell 1984b); the response may be viewed as an adaptive one; inhibition of forward movement and of ventilation would prevent further intake of the toxicant, and bradycardia and hypotension would diminish the transport of the toxicant from the gills to the tissues. The reflex may also have a role in guarding against gill oedema; the heart, when driven by increased venous return resulting from vigorous muscle pump activity, can, it seems, be reflexly inhibited from the gills. The branchial nerves and vagus provide the afferent pathway for the reflex; the cardiac vagus is the efferent pathway.

Whilst the receptors in the dogfish gill have been shown to be on or beneath the gill epithelium, and to be discharged by 5HT, they have not been more precisely identified. A study by Dunel-Erb *et al.* (1982) on neuroepithelial cells in the gills of perch, trout, eel and dogfish may be relevant. Neuroepithelial cells have been visualized; they are located beneath the epithelium of the filaments and are innervated with fine unmyelinated nerve fibres with which they make synaptic contacts. They are clustered together to form so-called neuroepithelial bodies and contain many dense cored vesicles indicative of stored catecholamine. They show a bright fluorescence identified as being that of 5HT. Such innervated epithelial bodies are known to occur widely in the airways of the lungs of air-breathing vertebrates and are regarded as being members of the APUD system. This is a system of scattered neuroendocrine cells

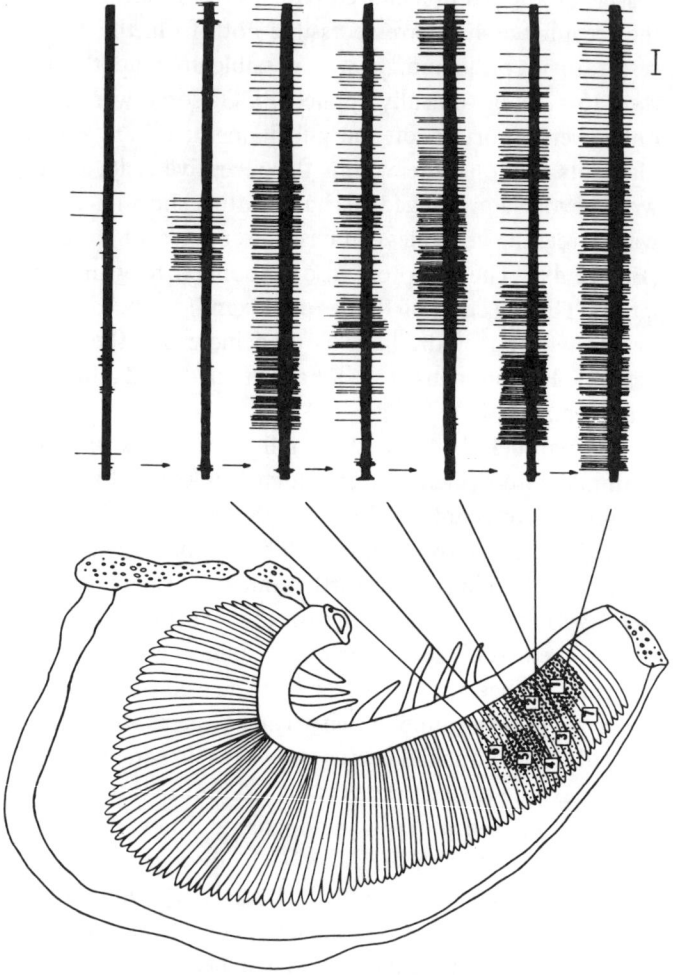

Figure 49. The receptive field of a gill nocioceptor of the spiny dogfish. The discharges in its afferent nerve fibre are shown from six different sites in the receptive field, in response to localized chemical stimulation with phenyldiguanide. The top trace shows the background discharge. The arrows mark the times at which the PDG was applied. Time marker = 1 sec. Redrawn from Satchell 1978.

which occur in various tissues of the vertebrate body and which have certain features in common. The acronym APUD (= amine precursor uptake decarboxylation) indicates their ability to take up amino acids and synthesize catecholamine; the system was first described by Pearse (1969). Many are believed to be receptors and when appropriately stimulated, catecholamine is liberated at the synapses and the afferent nerve terminals are excited. The application of 5HT to the gill epithelium would be likely to discharge nociceptors, for it would bypass the neuroepithelial cell and directly depolarize the nerve terminals.

10

THE RESPONSE TO EXERCISE

Introduction

In Chapter 6 the location and circulation of the red and white myotomal muscles were described; red muscle, it was noted, was aerobic and provided for sustained swimming, whereas white muscle was poorly circulated, depended largely on glycolysis and was used only in short bursts, for fast swimming. It is necessary to define these terms more precisely.

The speed of swimming, U, is related to the length L of the fish; Thompson (1917) and later Brett (1965) concluded that it was proportional to the length of the fish raised to the 0.5 power. U is expressed in $cm\ sec^{-1}$; $L\ sec^{-1}$ is also used. Swimming is arbitrarily defined as sustained, prolonged, or burst. Sustained swimming has to be able to be maintained for long periods, i.e. more than 200 min, without fatigue and is the swimming used when fish forage, school, migrate, or, as in the tunas, which lack a swimbladder, have to swim fast enough to counteract their negative buoyancy. Sustained swimming speeds vary, depending on fish length; Beamish (1978) gives values, for Atlantic salmon $12–25\ cm\ sec^{-1}$, for trout $0.2–26\ cm\ sec^{-1}$, for pike $5.5\ cm\ sec^{-1}$.

Prolonged swimming is observed over shorter periods, 20–200 min, and ends in fatigue. Such speeds are most frequently measured in laboratory experiments, when fish are swum in flumes and raceways. The symbol U_{crit} is used to indicate the maximum speed that can be maintained over a given period, often 1 h. U_{crit} for Atlantic salmon is $70–100\ cm\ sec^{-1}$, for trout $48–70\ cm\ sec^{-1}$, for pike $148\ cm\ sec^{-1}$. In different species a certain fraction of U_{crit} can be sustained purely by aerobic metabolism, whereas levels above this cannot. In trout this fraction is 80–90%; oxygen uptake increases exponentially with speed, and environmental hypoxia, or anaemia, which diminish oxygen delivery to the red muscle, lower U_{crit}.

Burst swimming is a rapid forward movement, involving a short period of acceleration and a sprint, and in total does not last longer than 20 sec. It is important in the life of many fish and is used to escape predators, capture prey, and pass through rapids in upstream migrations to spawning grounds. Ascertained burst speeds for Atlantic salmon are 429–600 cm sec^{-1}, for trout 186–226 cm sec^{-1}, for pike 360–450 cm sec^{-1}. Often these rapid dashes last only for a second or so. The velocities for the three categories of swimming quoted above are extracted from Beamish (1978), and have been collected from the literature. They do not refer to fish of the same body length, but nevertheless give an idea of the range involved. During the brief bursts performed at the highest speeds ventilation is halted as in higher vertebrates.

In the previous Chapter the possibility was mentioned that some of the adjustments of the heart and blood vessels, necessary to match the circulatory system to the various demands of these three levels of swimming, might be achieved purely by intrinsic regulation. The extent of these changes and the role of intrinsic mechanisms and of changes mediated by the autonomic nervous system, require examination.

Exercise leads to an increase in V_{O_2}, the oxygen consumption, and in salmonid fish, ventilation Vg, may increase by 5–15 times. This increase calls for integrated adjustments both to Vg and to circulation. In the trout studied by Kiceniuk and Jones (1977), Vg increased linearly from 200 to 1700 ml kg^{-1} min^{-1} in the transition from rest to U_{crit}; and V_{O_2} increased from 25 to 194 μmol O$_2$ kg^{-1} min^{-1}. We might expect that the circulatory changes likely to accompany this will involve an increase in cardiac output, and thus a greater flow of blood through the gills to transport their greater uptake of oxygen, an increase in ventral and dorsal aortic blood pressures, and a redistribution of blood away from inactive tissues to the red muscle.

The response of the heart

The extent of the increase in cardiac output

In the trout studied by Kiceniuk and Jones (1977) the output of the heart increased threefold, from 17.6 ml kg^{-1} min^{-1} to 52.6 ml kg^{-1} min^{-1} at U_{crit}. This was achieved by a rise in heart rate from 38–51 bpm, an increase of $\times 1.36$, and a rise in stroke volume from 0.46 ml kg^{-1} to 1.03 ml kg^{-1}, an increase of $\times 2.24$. A. P. Farrell (personal communication) gives the two data, heart rate HR, and stroke volume SV, and the amount of increase of each between rest and maximum exercise, for a number of species, e.g. Atlantic hagfish: HR 22–26 = $\times 1.18$, SV 0.4–

0.85 = × 2.07; sea raven: HR 45–53 = × 1.17, SV 0.3–0.47 = × 1.56; chinook salmon: HR 30–40 = × 1.33, SV 0.48–1.12 = × 2.3. The sea raven has the least increase of heart rate and stroke volume, yielding a rise in cardiac output of × 1.8; the two salmonids have the greatest increases of cardiac output, of × 3. How is this increase achieved? It might be, at least in part, an intrinsic response; it might result from changes evoked from either vagal parasympathetic or spinal sympathetic nerves; it might result from centrally liberated catecholamines.

The intrinsic mechanism: the Starling response

We have, in Chapter 2, noted the studies of Farrell (1984) and Farrell *et al.* (1982, 1983, 1985, 1986b) on the trout and sea raven, in which the heart, *in situ* in the pericardium, was perfused at different input pressures (preloads). They showed that whilst cardiac output and stroke volume increased greatly, there was no increase in heart rate. Similar findings are reported by Stuart *et al.* (1983) for the perfused heart of the buffalo sculpin (*Enophrys bison*). It is thus not likely that the increase in heart rate seen during swimming at speeds above 50% of U_{crit} is an intrinsic response to increased venous return. But the rise in stroke volume may be; the heart is very sensitive to small increases in the pressure of the returning blood (Farrell *et al.* 1985), and a rise of 1 mm Hg increased cardiac output in the sea raven fourfold. Farrell (1984) reported that in trout at U_{crit}, central venous pressure increased by 0.5 mm Hg, an increment almost sufficient to account for the increase in cardiac output actually measured *in vivo*. We also noted, in Chapter 2, that increased afterload resulted in homeometric regulation; ventral aortic pressure does indeed increase during swimming and this intrinsic response also must elevate cardiac work.

The hagfish heart, unlike those of elasmobranchs and teleosts, speeds as perfusion pressure is increased, a response that perhaps reflects its lack of innervation. Nevertheless, there is only an 18% increase in rate, compared with a 100% increase in stroke volume (Axelsson *et al.* 1990).

In elasmobranchs, the heart has been reported not to increase its rate during swimming; Johansen *et al.* (1966) made telemetric recordings from swimming horn sharks which show the rate unchanged. The rise in cardiac output was due, the authors suggest, solely to an increase in stroke volume.

The role of the cardiac vagus

The heart of resting elasmobranchs has long been known to be under vagal restraint and atropine, or vagotomy increases the rate by 30%. In cod the resting rate of 30.5 increases to 42.7 beats min^{-1} following exercise. Atropine significantly increases the resting heart rate (from 30 to 44 beats min^{-1}) and Axelsson (1988) calculates that cholinergic tonus accounts for 38% at rest and 15% following exercise. Axelsson *et al.* (1987) have shown small but significant cholinergic and adrenergic influences on the hearts of resting sea raven and Stevens *et al.* (1972) found that atropine increased the heart rate of resting lingcod, suggesting a resting vagal tone in this teleost.

In contrast, Wood and Shelton (1980b) found atropine produced no change in the heart rate of trout at 15 °C, although it blocked the occasional runs of spontaneous bradycardia. Cholinergic control of the heart decreases with rise in temperature. Priede (1974) found that trout swimming at speeds up to 0.6 L sec^{-1} did not increase their heart rate, but above this speed, rate increased, from 25–35 at rest to 56 beats min^{-1} at 1.07 L sec^{-1}; at yet higher speeds no further increase in rate occurred. There are probably considerable differences in the cholinergic tonus to the heart between species; clearly, when the heart is not subject to a tonic vagal inhibition, a reduction in vagal restraint is not a consideration. Nevertheless, it seems likely that vagal release does play an important part in rate increase.

The role of circulating catecholamines

Stress of any kind causes the level of catecholamines in the blood to rise, including the minor stress of moving the fish from a holding tank into the laboratory, and even more, that of the surgery necessary for placing cannulae. With ever greater care taken over the last decade to minimize this source of error, the values, in successive publications, for the 'resting levels' of noradrenaline and adrenaline have declined. Tetens and Christensen (1987) give the value for trout of noradrenaline 1.23 ± 0.09 and of adrenaline 0.63 ± 0.09 nmol l^{-1}. Other acceptably low values for trout are those of Primmett *et al.* (1986): noradrenaline $0.74 + 0.1$, adrenaline $0.91 + 0.1$ nmol l^{-1}.

Sustained swimming (1 length sec^{-1}) changes these levels rather little (noradrenaline 2.5 ± 2.0, adrenaline 0.3 ± 0.7 nmol l^{-1}) (Butler *et al.* 1986). Axelsson and Nilsson (1986), in their study of cod, showed that there is no increase in circulating catecholamines up to swimming speeds of 2–3 L sec^{-1}. But speeds above this, and particularly burst swimming, caused catecholamines to rise greatly; in trout repeated burst swimming raised blood catecholamines to: noradrenaline 85 ± 46, adrenaline 212 ± 89 nmol l^{-1}. In the spotted dogfish Butler *et al.* (1986) report that at rest plasma catecholamines were at concentrations of 10^{-9}–10^{-10} mol l^{-1}. But in this species, and in other elasmobranchs, catecholamines are secreted even during spontaneous swimming; in the dogfish noradrenaline rose 2.3 times to 32 and adrenaline 3.3 times to 19 nmol l^{-1}. Fish certainly have the mechanisms to liberate catecholamines and Holmgren (1977) has shown in the cod that electrical stimulation of the preganglionic fibres leading to the chromaffin tissue in the pronephros causes a release of sufficient catecholamine to affect the heart.

Catecholamines have diverse effects on the fish heart. They increase the intrinsic rate only by 15–20% compared with the 100% increase reported in some mammals. This is presumably related to the fact that changes in ventricular volume, in the single-ventricled amphibians and reptiles, may lead to undesirable mixing of the arterial and venous streams of blood, as we noted in Chapter 2. Graham and Farrell (1989) present evidence that the very low levels of catecholamines noted above as present in resting fish tonically maintain the excitability of the A–V node and, when absent, conduction may fail and arrhythmias occur. Catecholamines enhance the Starling response so that larger increases in stroke volume and cardiac output occur for each unit of increased filling pressure. Farrell *et al.* (1986b) conclude that adrenaline, at physiological concentrations, increases the sensitivity of the *in situ* perfused trout heart to preload by 45–85%. The response is mediated by β_2 receptors. Catecholamines improve the contractility of the acidotic heart, perhaps by increasing the flow of Ca^{+2} through the sarcolemma, and this may assist in the recovery after exercise (Farrell 1985).

The role of cardioaccelerator nerves

The fact remains, as we noted above, that cardiac acceleration does occur in teleosts when swimming at speeds above 50% U_{crit}; it may be related to the liberation of catecholamines from the chromaffin tissue but it also may be caused by the release of these agonists from the terminals of cardioaccelerator nerves to the pace-maker of the heart. Axelsson (1988) has examined the effect of sotalol, a β blocking drug, and/or atropine on the heart rate of cod at rest and during exercise. The study suggests that adrenergic tonus, comprising the summed affects of circulating catecholamines and adrenergic innervation, accounts for 21% of the tonus at rest and 28% after exercise. The role of nerve-mediated versus centrally liberated catecholamines has also been examined by injecting the drug bretylium. It caused a small reduction in heart rate in both resting and exercising cod (Axelsson and Nilsson 1986), suggesting that adrenergic cardioaccelerator autonomic fibres are active both at rest and during exercise. It is likely that in many species, as in the cod, sympathetic excitation combines with vagal release to speed the heart.

The response of the gill vasculature
Intrinsic dilation of gill vessels
As all of the cardiac output passes through the gills, gill blood flow in trout increases threefold at 90–100% of U_{crit}. Mean ventral aortic pressure rises from a resting level of 39 mm Hg to 62 mm Hg and dorsal aortic pressure from 31 to 37 mm Hg (Kiceniuk and Jones 1977). We noted in Chapter 6 that as the afferent lamellar arteriole has a greater resistance than the efferent arteriole, blood pressure within the lamella approximates most closely to that in the dorsal aorta. A rise in intra-lamellar pressure causes the pillar cell flanges to bulge and increases the thickness of the intralamellar space, thereby reducing its vascular resist-ance. Moreover, increasing pressure and increasing pulse pressure cause more lamellae to be recruited as the critical closing pressures of their afferent arterioles are exceeded, and this recruitment tends to occur at the distal end of the filaments (Farrell 1980a, Farrell *et al.* 1980).

The combined effect of these intrinsic responses is to decrease the overall resistance of the gill circulation. Wood (1974a) reports that in trout an increase in the differential pressure across the gill of 7.4 mm Hg decreases gill vascular resistance, R_g, by 17%. His studies also suggest that raised pressure evokes an increase in the intrinsic tone of the vascular smooth muscle.

Parasympathetic nerve fibres to the gills
Early studies in elasmobranchs suggested that electrical stimulation of the peripheral ends of cut IX (glossopharyngeal) and vagal branchial nerves caused a rise in ventral aortic blood pressure, suggesting that vasocon-strictor fibres innervate gill vessels. Metcalf and Butler (1984) showed that in the spotted dogfish this is an artefact of the movements made by the gill filaments, and when their skeletal muscles are paralysed by the neuromu-scular blocking drug, pancuronium, these changes do not appear.

Branchial cholinergic vasoconstrictor fibres do, however, occur in teleosts. Isolated perfused trout gills have been studied by observing the distribution of Prussian blue particles before and after the addition of acetylcholine to the perfusate (Smith 1977). The vasoconstriction it pro-duced prevented most of the particles of dye from entering the secondary lamellae. Retrograde perfusion identified a significant constriction at the base of each efferent filamentar artery, i.e. in the region immediately prior to its entry into the efferent branchial vessel. This zone is known to be the site of a dense innervation of fine nerve fibres containing small clear vesicles characteristic of cholinergic terminals (Laurent and Dunel 1980). Electrical stimulation of branchial nerves in isolated perfused cod gills brought about an increase in resistance which was, in part, abolished by atropine. Moreover, microvascular pressure studies, made when acetylcholine was administered, showed that pressure in the efferent filamentar artery rose towars the level of that in the afferent vessel, whilst that in the efferent branchial artery fell almost to zero (Farrell and Smith 1981). As acetylcholine is never a circulating hormone, these studies suggest that each efferent filamentar artery possesses a site of nerve-mediated cholinergic vasoconstriction at its base. In trout and perch Dunel-Erb *et al.* (1989) have identified, in addition, a cholinergic innervation of the efferent lamellar arteriole and the proximal region of the efferent filamentar artery.

The response of the gill vessels to branchial efferent adrenergic fibre stimulation

The longitudinal chains of the autonomic nervous system extend into the head of teleosts and carry adrenergic fibres to the gills via the branchial nerves (Chapter 9). Nilsson and Pettersson (1981) electrically stimulated this nerve chain in a cod whilst a gill was perfused *in situ*. Such stimulation caused a rise in perfusion pressure, an increased flow from the efferent artery, and a decreased flow from the venous drainage of the gills. Donald (1984), on the basis of catecholamine fluorescence histochemistry, has identified adrenergic fibres innervating the vessels of the nutritive circulation (Chapter 6) in the trout gill. Vogel *et al.* (1974) have described, in tilapia, an accumulation of fine nerve terminals around the arteriovenous anastomoses, and Nilsson and Pettersson (1981) suggest that it is the constriction of these that accounts for the responses to electrical stimulation of the branchial efferent fibres. That a reduction of flow through the arteriovenous path would lead to an increase of flow through the arterioarterial pathway would be expected, for the two are in parallel. The arteriovenous anastomosis was suggested as the site of an α adrenergic constriction also by Payan and Girard (1977) who perfused the isolated head of the trout. The constriction may off-set the increased flow through the arteriovenous pathway which will occur when raised intralamellar pressure results from the cholinergic constriction of the efferent arterial outflow, mentioned above. The two opposing responses may mediate the demands for increased flows in the arterioarterial and arteriovenous pathways to service the conflicting interest of oxygen uptake and ionic regulation.

The response of the gill vessels to centrally liberated catecholamines

In a study of isolated perfused trout gills, Wood (1975) described their response to an injection of adrenaline AD, or noradrenaline NAd, as a fleeting vasoconstriction followed by a sustained dilation. The constriction is due to α receptors, the dilation to β receptors. Isoprenaline (ISO), a pure β agonist, caused only dilation. The sensitivity of the gill vessels to these amines was much greater than that of the peripheral vessels. The maximum dilation is similar for all three catecholamines, and the doses required were in the order ISO = 10 nM, AD and NAd = 100 nM. All three, at maximum dose, effected a 60% reduction in gill resistance, which reflects the increase in the number of channels, as afferent lamellar arterioles were dilated and additional lamellae recruited. This dilation of the arterioarterial pathway is certainly the response most frequently reported for the administration of catecholamines to the branchial circulation.

We noted in Chapter 3 that electrical stimulation of the nerves to the chromaffin tissue in the pronephros of the cod (Nilsson *et al.* 1976) caused a liberation of NAd and AD into the blood of the posterior cardinal vein, and in experiments in which the tissue had been previously loaded with ^{3}H-adrenaline, labelled amine appeared in the blood. In another study (Wahlqvist 1981), catecholamines were released from the chromaffin tissue by electrical stimulation of the preganglionic fibres, and the heart was replaced by a peristaltic pump. The blood was perfused through a gill where it again produced an initial constriction and a subsequent dilation. These responses were separately blocked by α and β blockers. The predominant conclusion of these and similar studies is that the central liberation of catecholamines plays the predominant part in dilating gill vessels and increasing flow through them. Cutting the sympathetic chains in the head, and thereby depriving the gills of their adrenergic nerve supply, does not alter the resting value of flow and resistance in cod, but cutting the nerve supply to the chromaffin tissue markedly reduces the tonus of the vessels at rest, and lessens their ability to respond to change (Wahlqvist and Nilsson 1980).

Although gill vesssels have the ability to respond to centrally liberated catecholamines in the ways outlined above, it needs to be reiterated that in those teleosts that have been most investigated, i.e. cod and trout, catecholamine levels do not rise greatly in moderate exercise. Such responses have probably greater importance during burst swimming, or when sustained swimming occurs in the presence of anaemia or hypoxia. The topic will be discussed further in Chapter 11.

The response of the peripheral vessels
Intrinsic responses

Studies of perfused tissues of higher vertebrates show that activity, which lowers P_{O_2} and causes the accumulation of metabolites such as H^+, CO_2 and lactate, brings about hyperaemia, i.e. a dilation of blood vessels and an increase in flow, because their smooth muscle is relaxed by these agents (Chapter 4). Canty and Farrell (1985) have studied this aspect of autoregulation in the perfused trunk of the ocean pout, a fish in which the myotomes consist almost entirely of white muscle. Changing the pH through the physiological range brought about a vasodilation with acidosis and a vasoconstriction with alkalosis. Lowering the pH from 7.9 to 7.4 increased flow by 8%; raising it to pH 8.5 diminished flow by 37%. There was no additional vasodilation below pH 7.4. In contrast to the condition in mammals, hypoxia had no effect and the lack of response to it is probably related to the tissues' small need for O_2. The authors suggest that species like trout and eel, with 4–16% of red muscle, may show an autoregulatory response to hypoxia as well.

Blood pressure rises during exercise, dilates vessels, and increases flow through them. But this increase is likely to be small; the rise in dorsal aortic pressure of trout at U_{crit} is only from 31 to 37 mm Hg (Kiceniuk and Jones 1977). The study of Wood and Shelton (1975) on the elasticity of the systemic vessels in the perfused trunk of trout shows that this would increase flow by only a few per cent.

The operation of the haemal arch pump (Chapter 8), and possibly other auxiliary pumping mechanisms associated with the contraction and relaxation of the myotomes, will assist the flow of blood through the trunk muscles, but again there is no reason to suppose that the red muscles would be favoured.

By injecting radioactive rubidium, and radio-labelled microspheres into exercising trout (Randall and Daxboeck 1982) it was possible to compare the flow, in ml $min^{-1}kg^{-1}$ of different tissues (Table 9).

It is clear that there is a quite remarkable channelling of the flow to the red muscle at the expense of the white muscle. The authors calculate that 93% of the increase in oxygen uptake is delivered to the red muscle and its oxygen utilization increased by a factor of 12. How, we may ask, is this redistribution brought about?

Table 9. *Calculation of muscle and systemic blood flow* $(ml\ min^{-1}\ kg^{-1})$
during rest and exercise in trout

	Rest	Exercise (80% U_{crit})
Red and white muscle (mixed)	5.5	10.2
White muscle	4.8	0.4
Red muscle in white muscle	0.7	9.7
Red muscle (lateral)	1.1	16.2
Rest of body	6.2	11.8
Total flow = cardiac output	12.8	38.2

The response to circulating catecholamines

We have already noted that catecholamine levels do not increase greatly during exercise, but it is necessary to examine the response to injected amines because these may mimic the effect of neural release. Wood and Shelton (1975) emphasize how very much less sensitive the peripheral circulation is to these agents compared with the branchial vessels. The sensitivity of the gill circulation to catecholamines is such that it covers most of the dose–response curve derived from studies on isolated perfused preparations, but that of the isolated perfused trunk of trout scarcely reaches the lower end of its appropriate curve. The response to injected adrenaline and noradrenaline is vasoconstrictor, indicating the presence of α receptors. Isoprenaline has no effect in trout (Wood and Shelton 1975), but causes vasodilation in the short-finned eel (*Anguilla australis schmidtii*) (Davie 1981b).

The role of the autonomic vasomotor nerves

The spleen of exercising yellow tail has been observed, through a window, to contract during severe exercise and the haemoglobin content of the circulating blood increases (Yamamoto *et al.* 1985). Nilsson and Grove (1974) report that catecholamines contract the spleen in the cod. In exercised carp Yamamoto and Itazawa (1989) have shown that the increase in haematocrit and haemoglobin content that occurs is not solely caused by haemoconcentration due to water loss, for plasma protein does not rise proportionately. Following severe exercise for 1 h, 33% of the erythrocytes in the blood had been expelled from the spleen. When cod swim at U_{crit}, haematocrit increases from 14.3 to 18.5; if the central chromaffin tissue is denervated the rise in haematocrit is smaller but is still significant, suggesting that some of the contraction of the spleen is mediated by autonomic fibres (Butler *et al.* 1989).

Peripheral blood vessels are known to have an extensive adrenergic innervation (Campbell 1970). In teleosts there are adrenergic fibres amongst the vascular smooth muscle throughout the gut (Nilsson 1984). Wahlqvist and Nilsson (1981) have investigated the role of vasomotor nerves by electrically stimulating the sympathetic chains in the isolated perfused trunk of the cod. This caused vasoconstriction; phentolamine, in doses known to produce α blockade, sometimes revealed a small vasodilation. They also injected the catecholamines AD, ISO, and phenylephrine, all of which constricted the trunk vasculature. ISO, in some preparations and in low dosage, sometimes produced vasodilation. Smith *et al.* (1985) used the same preparation and in addition pretreated the fish with bretylium. After sufficient time had elapsed for the catecholamine-release mechanisms to be completely blocked, electrical stimulation of the sympathetic chains failed to produce vasoconstriction, nor was phentolamine able to lower the perfusion pressure of the preparation any further.

Bretylium, administered to resting or swimming cod (Axelsson and Nilsson 1986) lowered their blood pressure, and when it was followed by phentolamine, the pressure fell no further. Had any of the tonus of the blood vessels depended on circulating catecholamines, it would have survived the administration of bretylium but would have been reduced by phentolamine. The study of Randall and Daxboeck (1982) showed that during exercise, blood flow to the liver and stomach of trout was reduced. The conclusion drawn from these studies, and those of Smith (1978), Wood (1974b) and Wood and Shelton (1975) on trout, is that during exercise, blood pressure is maintained by, and blood flow is directed to the red muscle by, autonomic nerve fibres which constrict resistance vessels in the viscera and perhaps in the white muscle. In other genera, centrally liberated catecholamines may become increasingly important as the exercise becomes more stressful and have important effects on the branchial vessels, and the red cell membrane, topics treated in the next Chapter. Butler *et al.* (1989) report that there is a significant reduction of swimming performance at U_{crit} in cod in which the chromaffin tissue in the pronephros has been denervated, but this loss is in part restored when blood catecholamines are returned back to the expected level by infusion.

In summary

Fish vary greatly in their capacity to endure prolonged exercise and we need to accept with caution conclusions based on the study of so few species. Tentatively we may say that, when swimming at speeds up to 80% U_{crit} the increase in cardiac output is due less to an increase in rate and more to an increase in stroke volume, and that this is an intrinsic mechanism dependent on the Starling response to an increased preload. The increase in rate may depend on increased activity of adrenergic cardioaccelerator nerves and a reduction in vagal restraint. In the gills the increase in oxygen uptake depends largely on the intrinsic response of the gill vasculature to the increase in blood pressure which extends flow through a wider proportion of each lamella and recruits additional lamellae. Adrenergic and cholinergic fibres in the branchial nerves may constrict the efferent filamentar arteries and the entrances into the arteriovenous anastomoses, and thereby regulate the conflicting claims of oxygen uptake and ion balance. Nerve-mediated constriction of the spleen enhances the oxygen capacity of the blood by pouring extra erythrocytes into it. In the peripheral circulation the intrinsic response of reactive hyperaemia may play an important part in increasing flow to the red muscle, but autonomic nerve fibres to the resistance vessels of the viscera are also of importance in this.

11

THE RESPONSE TO HYPOXIA

Introduction

The proportion of oxygen in the atmosphere, relative to nitrogen is, in almost all circumstances, close to 21%, and hypoxia is a hazard encountered rarely, except by those air-breathing vertebrates which dwell in or fly at high altitudes or burrow beneath the ground. Fish, in contrast, may be subject to oxygen pressures that vary widely, from water which is supersaturated, with a P_{O_2} above 155 mm Hg, down to levels where oxygen is virtually absent. Photosynthesis by green algae and submerged aquatic plants may liberate oxygen into solution and cause hyperoxygenation; accumulations of organic material may, in their decay, take up much or all of the oxygen that would otherwise be available. The P_{O_2} of shallow water in bays may decrease to 20 mm Hg by sunrise (Kerstens *et al.* 1979). Water from beneath winter ice, following the spring melt, often has a low oxygen content, as has water imprisoned below the thermocline. Even in normoxic water, ventilation may be temporarily impeded by eating and swallowing whole prey. Fish respond to these challenges with a variety of measures which maximize oxygen uptake and economize in its expenditure. We can group the changes that hypoxia evokes into three categories.

(1) Some are reflex responses set in train by the stimulation of oxygen receptors (Chapter 9). The increase in ventilation, the bradycardia of hypoxia, cardiorespiratory synchrony, and the constriction of peripheral blood vessels are examples which will need to be considered.

(2) There are intrinsic responses of the gill vessels, which bring about changes in the pattern of blood flow. Some are a direct response to the rise in blood pressure that occurs as a result of the vasoconstriction mentioned above. There is also, it is believed, an intrinsic constriction of certain gill vessels in response to the lower P_{O_2}.

(3) Stress of any kind is liable to activate the autonomic centres which regulate the secretion of the chromaffin tissue of the pronephros and lead to the liberation of catecholamines into the blood; hypoxia is itself a form of stress and results in a rise in blood catecholamines. These, circulating to the various tissues of the body, evoke a further group of responses. They have a specific protective effect on the heart muscle, the perform-ance of which would otherwise be adversely affected by the falling pH. They cause further changes in the pattern of blood flow in the gill and of its oxygen diffusive conductance, which together enhance its uptake of oxygen. They increase the efflux of H^+ from the gills. They stimulate an exchange of H^+ with Na^+ across the membrane of the red cell, thereby increasing its internal pH relative to that of the plasma. They bring about a rise in the level of blood glucose which, it is believed, is beneficial as glycolysis replaces aerobic metabolism. We need to consider these various points in greater detail.

I   Reflex responses (A) the heart
The bradycardia of hypoxia

Hypoxia evokes an increase in ventilation, and a reflex bradycardia. In flounder, acute hypoxia (P_{O_2} = 30 mm Hg) causes a doubling of venti-lation volume (Kerstens *et al.* 1979): in lingcod ventilation increases three times at a P_{O_2} of 60 mm Hg (Farrell and Daxboeck 1981). The change in heart rate is equally notable. Smith and Jones (1978) report that trout at 7 °C have a resting heart rate of 45 bpm in normoxic water (P_{O_2} = 150 mm Hg) and this falls to 26 bpm in water with a P_{O_2} of 30 mm Hg. The response is very rapid and suggests that the relevant chemoreceptors are at or close to the surface. We reviewed, in Chapter 9, the evidence that they lie on or in the anterior face of the dorsal part of the first gill arch in trout and that they are distinct from the receptors which mediate the respiratory response.

The bradycardia of hypoxia is a reflex mediated by the cholinergic fibres of the cardiac vagus and it is abolished by atropine and vagotomy. Butler and Taylor (1971) found, in the spotted dogfish, that hypoxia of rapid onset produced in initial intense bradycardia which diminished towards a stable level characteristic of the P_{O_2} of the inspired water. The reduction in rate brings about an increase in stroke volume, a result of the Starling response, but this is sometimes insufficient to maintain cardiac output. Farrell's (1982) study on short term hypoxia in the lingcod showed that whilst stroke volume doubled, output fell by 31% when P_{O_2} was below 45 mm Hg. But at oxygen tensions down to 70 mm Hg the heart rate and cardiac output remained virtually unchanged although, at this severity of hypoxia, almost all the increase of ventilation had occurred. Wood and Shelton (1980b) exposed trout to hypoxic water with a P_{O_2} of 50–65 mm Hg and found that the bradycardia was compensated by the increase in stroke volume so that cardiac output was unchanged or even increased slightly. The results of other studies are similar and it is generally concluded that the reduction in output, if it occurs, is not great at moderate levels of hypoxia.

The functional significance of the bradycardia of hypoxia has been much debated. Randall and Shelton (1965) suggested that it allows the blood to stay in the gill lamellae longer; cardiac cycles are of greater length and each ventricular ejection moves a volume of blood into the lamellae which displaces the blood that is there. As the P_{O_2} of the water falls, the gradient of oxygen tension driving oxygen into the blood is lessened, and the longer time would allow diffusion to proceed further towards completion. Moreover, bradycardia allows a longer residence time for the blood in the spaces within the spongy myocardium, and, by like argument, enables this to increase its uptake of oxygen from the hypoxic blood flowing there (Farrell 1984).

Cardiorespiratory synchrony

Hypoxia evokes a second cardiac reflex, termed cardiorespiratory synchrony, by which the heart beat becomes coupled to the respiratory cycle. Each heart beat tends to occur at a particular phase of the respiratory cycle, and in resting fish the reflex is sometimes seen in normoxic water (Satchell 1960). Lesser degrees of coupling give rise to repeated runs of cardiac cycles of changing length, and these can be computer-simulated by treating the problem as one in which a saw tooth oscillator, the pace-maker potential of the heart, is modulated by a sinusoidal wave form, i.e. the inhibition generated by the trains of action potentials in the cardiac vagal fibres (Satchell 1968). The heart tends to beat as the mouth opens, and if the flow of water into the pharynx is manually controlled by a solenoid valve, the heart beats each time the water is allowed to flow over the gills (Randall and Smith 1967). The level of hypoxia at which cardiorespiratory synchrony becomes apparent can be related to the oxygen dissociation curve of the species of fish; its onset coincides with the level at which the haemoglobin would commence to be significantly unsaturated. Cardiorespiratory synchrony too is abolished by atropine and vagotomy. Its functional role may be to synchronize two pulsatile flows, i.e. blood and water, so that the period of rapid flow of the one, caused by ventricular ejection, coincides with the period of rapid flow of the other, caused by the expulsive phase of ventilation (Satchell 1960). It may also play a role in venous return in that the expulsion of blood from the anterior cardinal sinus by the branchial venous pump (Chapter 8) is coordinated with the aspiratory inflow of blood to the atrium.

Changes in coronary blood flow

We noted in Chapter 9 that the coronary vascular bed of fish, like that of higher vertebrates, contains both α constrictor and β_1 dilator adrenocep-tors (Davie and Daxboeck 1984, Daxboeck and Davie 1986), and, as in mammalian coronary vessels, adrenaline is a more potent α agonist than is noradrenaline. Ask (1983) has shown that in trout the dominant nerve terminal catecholamine is adrenaline, and nerve-mediated activity might be expected to cause constriction of the coronary vessels. However, the concomitant adrenergic nerve activity to the myocardium would cause it to be increasingly active, with the possibility that the coronary vessels dilate in response to local metabolites. The part played by the autonomic system in the control of the coronary circulation in fish remains uncertain.

I Reflex responses (B) The peripheral circulation

In the previous chapter it was noted that blood flow through peripheral vessels, both at rest and in moderate exercise, was primarily regulated by vasomotor nerves to the trunk vessels (Wood and Shelton 1975), and that these vessels are much less sensitive to circulating catecholamines than those of the gills. Hypoxia, in contrast, increases blood catecholamines to a level at which they may be of importance. Because α receptors predomi-nate in the trunk vasculature, resistance is minimal with the lowest concentrations of catecholamines and rises during hypoxia. Farrell's (1982) study of lingcod shows that pressure: flow relationships are steeper during hypoxia and the resistance of the peripheral circulation increases in water of P_{O_2} 25–45 mm Hg. The extent to which this is nerve-mediated or is due to the raised level of catecholamines is not at present clear; the levels of these present in the stressed cod noted above are sufficient to evoke vasoconstriction when compared with the levels sufficient to do so in the isolated perfused trunk (Wahlqvist 1980).

II The intrinsic responses of the gill vessels

The study, by Farrell and Daxboeck (1981), of oxygen uptake in lingcod showed that with progressive hypoxia, uptake was maintained down to a water oxygen tension of 45 mm Hg, an adjustment that entailed a four-

fold increase in ventilation. The winter flounder (Cech *et al.* 1977) maintained its uptake even when oxygen was reduced by 54%. Some of the
changes in the pattern of gill blood flow depend on intrinsic responses of
the lamellar vessels.

Farrell (1982) reports that in lingcod, moderate hypoxia (P_{O_2}
70–80 mm Hg) causes an increase in ventral and dorsal aortic pressure;
that in the dorsal aorta rose from 38.5 to 41.1 mm Hg. Holeton and
Randall (1967), in a study of trout, noted that systolic and pulse pressure
in both ventral and dorsal aortae increased steadily down to a P_{O_2} of
40 mm Hg, and only then showed signs of decreasing. The increase in
pulse pressure is directly due to the larger stroke volume, as bradycardia
lengthens the cardiac cycles.

We noted, in Chapter 6, that in the isolated perfused gill, the afferent
lamellar arterioles were the major site of gill resistance, a conclusion
confirmed in a morphometric study of lingcod gills (Farrell 1980b). An
increase in afferent branchial artery pressure and of pulse pressure caused
additional lamellae to be recruited (Farrell 1980a, Farrell *et al.* 1979).
Wood (1974a), in his study of perfused trout gills, noted that the branchial vasculature was very distensible and that a rise of differential pressure
from 26 to 41 mm Hg halved the resistance of the gills.

Farrell (1980b) reports, from the evidence of vascular casting, that
lamellar vascular pathways occupy almost 90% of the surface area. He
demonstrated that a rise in intralamellar transmural pressure caused an
increase in the thickness of the blood sheet within it. This distributed
blood more widely and particularly away from the basal marginal channel
towards the middle region of the lamella. This increase in the area
available for gas exhange in individual lamellae, and the recruitment of
additional lamellae, occurred in the isolated gill between oxygen partial
pressures of 44 and 22 mm Hg, which are within the physiological range
of the lingcod.

These two mechanisms may be augmented by an intrinsic vasoconstriction in response to hypoxia, of the efferent lamellar arteriole. Pettersson
and Johansen (1982) studied the responses in the isolated cod head, of
gills which could be irrigated with hypoxic water and/or perfused with
hypoxic saline. A vasoconstriction occurred whether the hypoxic stimulus
was delivered from outside or inside the lamella. The authors concluded
that the only site which could 'see' both stimuli was the efferent lamellar
arteriole. 'External hypoxia', they point out, is the physiologically significant hypoxia for fish and could be rapidly screened by a sphincter of
O_2-sensitive smooth muscle at this site. Ristori and Laurent (1977)

reported an increase in resistance of hypoxic perfused trout gills as did Satchell (1962) in the gills of the spiny dogfish.

III   The central liberation of catecholamines

There are now several studies which show that during hypoxia the level of catecholamines in the blood increases. In the spotted dogfish, exposure to water of P_{O_2} 35 mm Hg caused adrenaline to rise from 25.6 to 284.4 nmol l^{-1} and noradrenaline from 32.1 to 446.5 nmol l^{-1} (Butler *et al.* 1978). In the cod (Wahlqvist and Nilsson 1980) they rose from a resting level of 28.8 nmol l^{-1} of adrenaline and 13.2 nmol l^{-1} of noradrenaline to 292.5 and 32.1 nmol l^{-1}, respectively. We noted, in Chapter 10, that in the cod, the increase is almost all derived from the chromaffin tissue in the pronephros and when spinal nerve roots 1–4 are cut, electrical stimulation no longer causes a rise in catecholamine levels.

Hypoxia inevitably results in glycolysis and acidaemia. Recent studies on trout, in which acid is infused directly into the blood system, show that the release of adrenaline is directly proportional to the magnitude of the acidosis (Tang and Boutilier 1988). Boutilier *et al.* (1986) suggest that the increase is related to the rate of fall of pH, rather than its absolute value.

Circulating catecholamines have the potential to mediate changes in many systems; more and more of these changes are coming to notice each year. We will examine five of them.

The defence of the acidotic myocardium

We have already seen that, although the heart is reflexly slowed by hypoxia, cardiac output may change rather little. The stimulatory effects on rate and stroke volume which catecholamines evoke in isolated perfused hearts have been discussed in Chapters 2 and 10 but, *in vivo*, these tend to be overridden by nerve-mediated changes. Adrenaline is, however, also important in protecting the heart against extracellular acidosis. There are studies which show that stressful exercise can cause the pH of the blood to fall from around 7.9 to 7.4. This is harmful to the performance of the heart for, it is believed, H^+ competes with Ca^{2+} for binding sites on the troponin molecules of the muscle fibres (Gesser 1977); this reduces the force of contraction and lowers cardiac output. It is necessary, it seems, for the concentration of intracellular Ca^{2+} to be increased by some means if cardiac output is to be restored. Increased concentration of extracellular Ca^{2+} increases the force of contraction in the normal heart and can double it in acidotic strips of the ventricle of the sea raven (Farrell *et al.* 1988b).

Fish differ in the way they increase intracellular myocardial Ca^{2+} during hypoxia; species such as eel, flounder and plaice can recover contractile force by an intrinsic response in which extra calcium is made available within the myocytes, possibly from stores in the mictochondria (Carafoli 1985). But other species such as the sea raven, carp, trout, and cod do not show this recovery of contractile force. The study of the *in situ* perfused heart of the sea raven by Farrell *et al.* (1986b) showed that reducing the pH of the perfusate from 7.6 to 7.4 depressed cardiac output by 10–20% and the output remained depressed. It could be restored to the base level by adrenaline in physiological concentrations (Farrell 1985). Moreover, catecholamine has a similar restorative effect on a heart made acidotic by lowering plasma bicarbonate as it does on one subject to a respiratory acidosis (Farrell *et al.* 1988b).

Catecholamines do not act, as was formerly thought, by restoring the intracellular pH of the myocytes back to its preacidotic level, in the way that they do in red blood cells. Farrell and Milligan (1986) report that the intracellular pH in trout hearts fell from 7.53 to 7.25 when the extracellular pH was lowered from 8 to 7.45 and this change was not altered by adrenaline. How adrenaline restores the contractile force of the acidotic heart is uncertain. It may cause an influx of Ca^{2+} through the sarcolemma by opening calcium channels.

Catecholamines and the increase in the oxygen diffusive conductance of the gill

Perry *et al.* (1985), using the isolated, saline-perfused trout head, report that when the gills were ventilated with hyperoxic water ($P_{O_2} = 250$ mm Hg), increasing the rate of perfusion caused a reduction in the P_{O_2} of the dorsal aortic perfusate. This indicates that there was a limitation in the rate at which oxygen could diffuse into the perfusate as it passed through the gill lamellae. Adrenaline increased oxygen uptake into the perfusate from 0.44 to 0.49 μmol min^{-1} dl^{-1}. This stimulation of oxygen uptake by catecholamines was shown to be a β effect.

There is still much to be learnt about how this increase in the oxygen diffusive conductance of the gill occurs. Haywood *et al.* (1977) and Isaia *et al.* (1979) present evidence that adrenaline increases the permeability of membranes to small lipophilic and water soluble molecules such as oxygen and thereby enhances uptake across the lamellar wall. But other changes also occur. Booth (1979b) working on trout, and Holbert *et al.* (1979) working on the channel catfish, report that catecholamines increase the area of gill perfused with blood by recruiting additional lamellae, i.e. they augment the intrinsic response to increased blood pressure noted in (2) above. In the trout, injected catecholamines can reduce branchial vascular resistance by 60% (Wood, 1974a). Nekvasil and Olson (1986) report that 50% of a pulse of noradrenaline is removed as it passes through the gill vessels, presumably by binding to catecholamine receptors.

Pettersson (1983) concluded that this was a β response dependent on the dilation of afferent lamellar arterioles. It has to be emphasized, however, that although dilation and a reduction in resistance are evoked by the administration of catecholamines to isolated perfused gills, hypoxia causes a rise in gill resistance (Wahlqvist and Nilsson 1980). This, Pettersson (1983) suggests, is because catecholamines also exert an α action on the efferent lamellar arterioles which reinforces their intrinsic constrictor response to hypoxia. We have, in Chapter 10, noted that α constriction, in addition, curtails the flow into the arteriovenous anastomoses and favours flow through the arterioarterial rather than the arteriovenous pathway. There is evidence that in teleosts this outflow is also under the control of adrenergic nerve fibres travelling in the branchial nerves. However, their significance is obscure; Wahlqvist and Nilsson (1980) found that cutting the sympathetic nerve connectives in the cod head, and thereby depriving the gills of their adrenergic vasomotor fibres, did not lower the dorsal aortic pressure of stressed fish, and concluded that sympathetic fibres do not appear to be of importance in their response to hypoxia. In the spotted dogfish Metcalf and Butler (1988) found that α and β blockade did not diminish the ability of the gills to transfer oxygen to the blood during hypoxia.

The role of catecholamines in regulating oxygen uptake by erythrocytes
Recent work shows that the acidaemia that results when trout are exposed to a severe hypoxia of rapid onset, i.e. down to 15 mm Hg, occurs in two phases (Claireaux *et al.* 1988, Fievet *et al.* 1987, 1988). Phase 1 is of rapid onset, and of brief duration, for it lasts for only a few minutes. It is due to the release of H^+ from within the red blood cells. Catecholamines exert a β action on the red cell membrane, and acting via cyclic AMP, switch on an $Na^+–H^+$ exchange mechanism, or antiporter. The outflow of H^+ acidifies the plasma and alkalizes the cell interior. The mechanism is not related to the level of lactate in the blood, and is blocked by β antagonists. It is activated by the rise in blood catecholamines released from the chromaffin tissue. Phase 2 acidification, in contrast, is of slower onset, occurs in response even to mild grades of hypoxia, is directly associated

with the presence of lactate, and is not blocked by β antagonists. It is a consequence of the switch to glycolysis and the release of lactate from the skeletal muscles.

The study of the Phase 1 mechanism has been extended to blood *in vitro*. When catecholamines are added to suspensions of trout red blood cells there is an acidification of the medium, and an alkalinization of the cell interior (Baroin *et al.* 1984a, b, Cossins and Richardson 1985). The pH disequilibrium in turn activates the $Cl^--HCO_3^-$ exchange mechanism resident in the Band 3 protein of the cell membrane. The combined inflows of Na^+ and Cl^- cause the cells to swell by osmosis (Chiocchia and Motais 1989. Ferguson *et al.* 1989). The increase in cell volume in trout may amount to 15% (Nikinmaa and Huetis 1984). Cell swelling and the acidification of the blood do not occur if NaCl is replaced with choline chloride, or the antiporter is blocked by amiloride ($K_i \sim 10^{-4}$ mol l^{-1}). The rate of acidification is a saturable function of the extracellular Na^+ concentration. The capacity of the catecholamine-stimulated exchange mechanism is sensitive to medium pH and has a peak at pH 7.24. In trout it is highest during the summer and, in Britain, declines during November to March (Cossins and Kilbey 1989).

These changes counteract the reduction in the amount of oxygen carried by the blood, as the P_{O_2} falls, through two separate mechanisms. Despite the falling pH of the plasma, the pH of the interior of the red cell is maintained and the reduction of oxygen binding that would occur due to the Bohr shift is off-set (Primmett *et al.* 1986, Perry and Vermette 1987). In a buffered saline, adrenaline increased the intracellular pH and caused a 22–46% increase in the oxygenation of trout red cells (Nikinmaa 1983). Without this mechanism, the acidification of the plasma by glycolysis (Phase 2) would diminish the blood's ability to transport oxygen at the time it was most needed. The stimulation of the Na^+-H^+ antiporter by catecholamines, we noted, lasts for only a few minutes but the internal pH of the red cells is sustained throughout the duration of the hypoxia by some other mechanism that is not as yet understood (Fievet *et al.* 1988). Elasmobranch red cells appear to lack the β receptor and in the big skate (*Raja binoculata*) and Pacific dogfish (*Squalus suckleyi*), Tufts and Randall (1989) find that stimulation of red blood cells with isoproternol does not change their internal pH.

We noted in Chapter 4 that cell swelling enhances oxygen binding by reducing the concentration of ATP and other phosphates (Ling and Wells 1985, Fuchs and Albers 1988). At pH 7.4 adrenaline lowered the concentration of ATP in trout cells by 35%. The decline in erythrocytic

ATP: Hb_4 molar ratios in trout with increasing hypoxic stress has the effect of sustaining haemoglobin oxygen affinity so that it does not change over a period of 24 h (Boutilier *et al.* 1988).

The increase in oxygen binding caused by cell swelling would be of little avail if their larger size increased the vascular resistance to their flow. We noted in Chapter 4 that this does not occur because catecholamines also increase the deformability of red cells as measured by their ability to pass through a Nucleopore filter (Hughes and Kikuchi 1984). Chiocchia and Motais (1989) suggest that phosphorylation of cytoskeleton proteins may account for this change.

Milligan and Wood (1987) calculate that the reduction in red cell NTP concentration is greater than can be accounted for solely by swelling and there must also be some metabolic degradation of it. Studies show that β stimulation lowers the NTP: Hb ratio but it is then maintained constant at the lower level as long as aerobic metabolism can provide for ATP turnover (Ferguson *et al.* 1989). Beta stimulation leads to an increase in red cell oxygen consumption associated with this and under hypoxic conditions the NTP: Hb ratio fails to be sustained. Adjustments to red cell energy metabolism are an important part of the mechanism of adrenergically mediated regulation of red cell pH. Trout subjected to β blockade after stressing exercise show a higher mortality (van Dijk and Wood 1988).

In vitro studies of trout blood (Tetens *et al.* 1988) show that stimulation of proton extrusion by the red cells must be mediated very largely by noradrenaline rather than adrenaline. The adrenoceptor affinity, as expressed in terms of the concentration of the catecholamine required to produce 50% of the maximum effect, the EC_{50}, is: adrenaline, $7.6 \pm 7.8 \times 10^{-7}$; noradrenaline, $1.3 \pm 0.6 \times 10^{-8}$ mol l^{-1}. Isoprenaline is even more potent. The potency order ISO: NAd: AD indicates that these are β_1 receptors. This contrasts with the red cells of amphibians and mammals where proton extrusion is mediated by β_2 receptors. When the concentrations of the two catecholamines in trout blood are related to their EC_{50} it is clear that the adrenoceptors of the erythrocytes must be binding noradrenaline almost exclusively.

Effects on the efflux of H^+ from the gill

Following exhaustive exercise there is a transient net efflux of H^+ from the gills. Recent studies (Tang *et al.* 1988) on trout show that when HCl is infused into sea water this too stimulates an efflux of H^+. This is reduced by the presence of a β blocker suggesting that it is evoked by catecholamines. Beta adrenergic stimulation of branchial Na^+/H^+ and Na^+/NH_4^+ exchanges, combined with an inhibition of Cl^-/HCO_3^- exchange, are believed to be chiefly responsible but the catecholamine-mediated general increase in permeability of membranes to small lipophilic molecules noted by Isaia *et al.* (1979) may also contribute.

Effects on carbohydrate metabolism

The hormone chiefly responsible for regulating the concentration of glucose in the blood, in fish as in other vertebrates, is insulin; increased secretion of insulin decreases glucose concentration. Countering the effect of this, are other hormones with a hyperglycemic action of which glucagon, adrenaline and cortisol are the most important. Adrenaline from the plasma binds to β receptors on the muscle membrane and the complex activates cyclic AMP (cyclic 3'5'-adenosine monophosphate). This nucleotide is synthesized within the membrane from ATP by an enzyme, adenylate cyclase, and β activation increases the activity of this. Cyclic AMP acts as a second messenger in the pathway and the final outcome within the muscle cell is that glucose molecules are split off from glycogen by phosphorylation and pass into the blood stream. Exposure of trout to water of P_{O_2} 30 mm Hg for 24 h decreased the glycogen of the white muscle by 50% (Boutilier *et al.* 1988).

The blood of spotted dogfish, withdrawn on day 3 following implantation of a cannula, has a plasma glucose concentration of some 41 mg%, i.e. 58% that of man (De Roos and De Roos 1978). Infusion of 75 μg kg^{-1} of adrenaline brings about an abrupt rise which increases this concentration by 72% in 3 h; levels remain significantly elevated for 12 h. Infusion of the same amount of noradrenaline increases glucose concentration only by 21%. In the starry dogfish 75 μg adrenaline cause an even larger increase of blood glucose, of 112% after 1 h. Thorpe and Ince (1974) report that in pike, 50 μg kg^{-1} of adrenaline increases blood glucose from 60 to 140 mg% in 30 min; the level then gradually returns to normal over a period of 5 h. Studies on carp liver show that in this tissue also, increase in blood glucose is primarily derived from stored glycogen; glycogenolysis occurs at a rate at least five and generally more than 30 times that of glyconeogenesis (Janssens and Waterman 1988). Ballatori and Boyer (1988) report that in isolated hepatocytes of the little skate, alanine is not a major gluconeogenic precursor as it is in higher vertebrates.

Butler *et al.* (1979) and Metcalf and Butler (1989) note that prolonged hypoxia causes a tenfold increase in catecholamines in the blood of spotted dogfish at 1.5 h. They suggest that the increase in blood glucose this causes may, in part, compensate for the switch from aerobic to glycolytic metabolism. Glycolysis yields much less energy per mole of substrate (2, as against 36 mol ATP per glucose) than does aerobic metabolism and an enhanced glucose level may help to compensate for this.

Hypoxia-induced changes in carbohydrate metabolism can, in turn, have effects on behaviour. During hypoxia lactic acid is oxidized to ethanol, which is neutral and easily eliminated through the gills. In goldfish, and presumably other species, ethanol carried in the blood to the brain activates the nucleus preopticus periventriculus, the behavioural thermoregulatory centre in the anterior brain stem, and causes them to seek cooler water; this lowers their metabolic rate and helps to conserve their limited store of glycogen (Crawshaw *et al.* 1989).

In summary

We know that in trout, hypoxia can be sensed by superficial receptors in the first gill arch and that their discharge reflexly evokes autonomic, vagally-relayed reflexes of bradycardia and cardiorespiratory synchrony. Both the slowing of the heart, and the coupling of its beat with the respiratory cycle, may assist oxygen uptake. In addition hypoxia causes vasoconstriction in peripheral vascular beds, and in consequence, an increase in their resistance, and a rise in blood pressure and pulse pressure. Intrinsic responses to this pressure increase in gill vessels enhance the area perfused with blood, and are supplemented by an intrinsic response to hypoxia of the efferent lamellar arteriole.

Hypoxia and the glycolysis that accompanies it increase the acidity of the blood, which in turn increases the liberation of catecholamines from the chromaffin tissue. These amines increase the branchial oxygen uptake which would otherwise be depressed by the low oxygen tension; the surface available for gas exchange is increased by recruitment of lamellae through the β mediated relaxation of afferent lamellar arterioles, and by the wider dispersal of flow within each lamella caused by the α mediated contraction of the efferent lamellar arterioles. Catecholamines enhance the efflux of H^+ from the gills. They restore the performance of the acidotic heart. They stimulate the exchange of intraerythrocytic H^+ with Na^+ thereby maintaining the oxygen affinity of the haemoglobin despite the increasing acidity of the plasma. They also increase the level of blood glucose which may be advantageous as glycolysis replaces aerobic metabolism.

The autonomic nervous system has a greater part to play in the response to hypoxia than it does in the response to exercise. This is perhaps not surprising; oxygen lack poses an immediate threat to the very tissues such as brain and retina which are needed to make an appropriate response to it; eye and brain are known to have poor glycolytic capacities. Hypoxia is a hazard which calls for an immediate adjustment of the action of the heart, the perfusion of the gills, the flow of blood to the tissues, and the content of oxygen and glucose carried in it. In contrast sustained swimming is involved in searching for food, in migrations associated with the reproductive cycle, and in social interactions such as seeking space and shelter. These are much more part of the normal life of the species and call for progressive increases in cardiac output and blood flow to the muscles; the readjustment of the circulatory system they involve are less radical than are those to hypoxia.

12

MYXINE, A SPECULATIVE CONCLUSION

Introduction

Myxine, the Atlantic hagfish, is the best known genus of a group of fish-like creatures, the myxinoids. Since the classical studies of Müller in the thirties of the last century, their status has puzzled biologists, for they combine many clearly primitive features with others which seem quite specialized. They are the only vertebrates in which the concentration of inorganic ions in the body fluids is sufficiently high to render them isosmotic with sea water, and they have no ability to osmoregulate. It is true that many elasmobranchs are isosmotic with sea water, but in such fish this is achieved by the retention of urea as 'osmotic stuffing'. Myxinoid haemoglobins are monomeric, with molecular weights around 17 500, and are thus more like circulating myoglobins than the tetrameric haemoglobins of fish. They lack true jaws derived, as these are in other vertebrates, from the gill arch of the second head segment, and their visceral arches are rudimentary and are not hinged to the skull but are direct extensions of it. Their gill lamellae occur inside sac-like pouches into and out of which water is pumped by encircling bands of muscle, an arrangement quite different from that of fish in which the whole pharynx is expanded and compressed by the hinged elements of the gill bars and the muscles that run between them. The alimentary canal is a straight tube and there is no true pancreas, for the islet cells are coalesced to form a little knot of tissue at the anterior end of the intestine, and the zymogen cells are scattered throughout the intestinal epithelium. Their insulin is one of the most highly substituted of all insulins, with 19 of its 51 amino acids different from those of a mammal such as the pig. They lack a spleen. They also lack a kidney in the sense that there is no separate organ or mass of tissue in which nephrons are grouped together, for a pair of large multilobed renal corpuscles occurs in each segment down the length of the animal, some 70 pairs in all, which open on each side, into an archinephric duct. The list could be continued, but it is clear that the myxinoids do indeed show some features which are plausibly ancestral to other vertebrates.

The myxinoids have, in earlier classifications, been united with the lampreys as the Cyclostomata. It has been increasingly realized in more recent studies that the two groups differ greatly; the body fluids of lampreys have an osmolarity approximately one third that of sea water, as do teleosts. Moreover, they can osmoregulate, and some genera migrate between rivers and the sea. There is a recognizable kidney, with the mesonephric tubules aggregated into a continuous structure. The pharynx is ventilated, in part, by a continuous layer of constrictor muscle; there are muscles in the gill septa and there are well developed cavernous bodies in the gills. Mallat (1984) likens them to elasmobranchs in respect of these and other features. Certainly, their circulatory systems are more like those of the true fish than they are of myxinoids. The heart has a well developed cartilagenous pericardium, and is vagally innervated. They have no caudal heart or portal heart. There is no subcutaneous sinus and the blood volume is only 8%. These differences have been discussed in detail by Hardisty (1982) and he locates the origin of the myxinoids in the early Cambrian, some 170 million years prior to that of the lampreys. The features which they do have in common, such as the monomeric haemoglobin, may reflect only that, like other archaic genera, the lampreys in the course of their evolution carried some primitive features on with them to a later date.

To some extent it is true that chance factors have distorted our view of the myxinoids, and resulted in an overemphasis on *Myxine* to the neglect of other genera. Its occurrence in the North Sea, around which most of the nineteenth century scientists were congregated, established it as the type for the group. The genus has some aberrant features. It buries itself, all but its head, in the mud at the bottom of fjords and emerges at intervals to feed. The water in which it lives changes little in temperature throughout the year and diurnal and seasonal changes in light intensity are minimal. It has no clearly defined diurnal activity rhythm, or seasonality of reproduction. *Eptatretus*, in contrast, is an active creature to be seen swimming in the shallow coastal waters of the Pacific. Like *Myxine* it will scavenge dead creatures from the sea bed but it is more actively predaceous and New Zealand hagfish present a problem to their investigators because of their tendency to attack each other in the aquarium tanks. Forster and his colleagues in Christchurch, New Zealand, have lately provided data on *E. cirrhatus*, the New Zealand hagfish, which has made possible a more complete view of the myxinoids. The questions may now be asked, if their circulatory system also shows features that can be regarded as primitive, and whether they can illuminate our understanding of the circulatory systems of the true fishes?

The heart
The performance of the myxinoid heart
The hearts of myxinoids are not small in relation to their size. The cardiac index of specimens of *Myxine* ranging from 48–176 g in weight was 0.18 (Satchell 1986); that of *Eptatretus* is 0.13 (M. E. Forster, unpublished). These values exceed 0.08, the mean for a series of teleosts, given by Poupa and Lindström (1983). Yet in some measured and calculated variables the values for hagfish hearts fall below those of elasmobranch and teleost fish. The myocardial power output of *Myxine* at rest has been variously determined as, 0.08 mW (Axelsson *et al.* 1990) and 0.05 mW per g ventricle (Driedzic *et al.* 1987). The maximum output of the perfused isolated ventricle of *Eptatretus cirrhatus* is 0.37 mW g^{-1} (Forster 1989). Even this maximum value is less than any of the four resting values for fish, given by Driedzic *et al.* (1987), i.e. spiny dogfish 0.64, sea raven 1.13, ocean pout 2.0, winter flounder 2.85 mW per g ventricle.

The mean ventral aortic blood pressure is 6.6 mm Hg (Satchell 1986), lower than that of any fish (Chapter 3, Table 4). That of *E. cirrhatus*, 10.8 mm Hg, is appreciably higher, reflecting its more active way of life (Forster *et al.* 1988, Davie *et al.* 1987). Nevertheless, the cardiac output of *Myxine* at rest is 8.9 ml kg^{-1} (Axelsson *et al.* 1990), a value not far below that of the spotted dogfish or eel (Chapter 2, Table 2), and this can increase by a factor of 2.5 during exercise, an increase larger than that of the cod or sea raven, and similar to that of a lingcod. The low myocardial power output of the hagfish heart arises, it is clear, not so much from a small cardiac output but from the very low pressure at which it is pumped.

How is the performance of the myxinoid heart to be explained? Certainly its gross and fine structure differs from that of the true fish in a number of respects.

The gross anatomy of the myxinoid heart
There are virtually no skeletal structures behind the skull, for the spine is replaced by an unrestricted notochord and there are no limb girdles and no paired fins. As a result, the pericardium is not attached to any skeletal

surfaces, as it is in most elasmobranchs and teleosts, and is thus unhindered in its ability to follow the changing volume of the heart. Moreover, the pericardium is not completely closed behind, and is widely in communication with the perivisceral coelom. As a consequence pericardial pressures are not subambient (Satchell 1986).

Perhaps related to the previous feature is the absence of a bulbus or conus arteriosus. The paired semilunar valves occur at the level of the junction of the rostral margin of the pericardium, with the ventricle (Grodzinski 1933). Immediately beyond this the ventral aorta extends rostrally and has an initial swollen region. This has been termed a bulbus by earlier authors but it is wrongly so named for it is not withdrawn into the pericardium, and does not bear the valves. Predictably, there is no B wave in the ECG (Figure 5*B*). The ventral aorta is very long as the heart is placed behind, not between the gills, and the elongate pharynx lies between it and the mouth,

Some features of the myxinoid cardiac cycle

Jensen (1965), in a microelectrode study of the atrial and ventricular fibres of the Pacific hagfish, drew attention to their rather low level of resting membrane potential, i.e. atrium 41 mV and ventricle 48 mV, values which lie below those given for carp, and yellowfin croaker in Chapter 2. This may reflect, as he suggests, the rather high level of plasma K^+ that characterizes the myxinoids. The amplitude of the ventricular action potential, 74 mV, involves an overshoot of 25 mV; this also is a value somewhat below those established for teleosts (Chapter 2).

The atrium is set to the left of the ventricle (Figure 3*A*) and is separated from it by a funnel-like connection (Johansen 1963). This perhaps explains a peculiar feature mentioned in Chapter 2, that the P wave of the ECG (Figure 5*B*) exhibits both a Pa and a Pt wave (Arlock 1975, Davie *et al.* 1987). Pt indicates the repolarization of the atrium, an electrical event the signal of which is normally hidden by the much larger QRS wave. The prolonged P–R interval reflects the greater distance and presumably, slow conduction velocity of the pathway linking atrium and ventricle. Systolic atrial pressure in *Myxine* (Satchell 1986) was 0.66 mm Hg and diastolic ventricular pressure was 0.59 mm Hg; there is only a small difference in pressure with which to fill the ventricle and longer times are needed for flow to occur.

The QRS wave of the ECG is of long duration, compared with that of a teleost. That of *Myxine* at 10 °C is 0.33 sec compared with 0.15 sec for the flounder at 5 °C (Biörck and Johansson 1955). This indicates that depolarization travels slowly through the ventricle; we noted in Chapter 3 that near-synchronous activation of the ventricular muscle is a prerequisite for generating high pressure. It is not possible, at present, to say whether this slow rate of activation of the myocardium is due to a paucity of gap junctions between the myocytes. Certainly, the rate of rise of ventricular pressure, dP/dT, is low. That of *Myxine* at 10 °C is 22 mm Hg sec^{-1} which can be compared with the value of 274 mm Hg sec^{-1} for the heart of lingcod at 15 °C (Stevens *et al.* 1972).

The heart of myxinoids, we have noted in Chapter 9, lacks autonomic innervation, and acetylcholine can be dripped onto it without altering its rate of beat. It is very evident, during laboratory investigations of live hagfish, that the heart is free of transient slowing if the animal is disturbed, the startle reflex, which is such a notable feature of elasmobranchs and teleosts. This lack of an ability to effect a beat by beat regulation explains also the absence of cardiorespiratory synchrony.

There are three features of the myxinoid heart in which it differs from that of the true fish, which are incompletely understood. As we noted in the previous Chapter, decreasing extracellular Ca^{2+} decreases the force of contraction in normal hearts, and can halve it in acidotic strips of ventricle. The heart of *Myxine* is little influenced by Ca^{2+} deprivation and even after 30 min at 8 °C, T_{max} stabilizes at about 50% of control (Lagerstrand *et al.* 1983, Poupa *et al.* 1984, 1985). The phenomenon has been termed the calcium paradox; the external lamina of the hagfish myocardium may, it has been suggested, have a higher Ca^{2+} binding capacity.

The heart is also rich in catecholamines, but early studies on isolated hearts suggested that they responded minimally to administered catecholamines. More recent research on intact unanaesthetized *Myxine* in which 10 nmol kg^{-1} adrenaline was perfused into the caudal vein, showed that this caused cardiac output to rise from 12 to 22 ml min^{-1} kg^{-1}. This resulted from increases in both rate and stroke volume (Axelsson *et al.* 1990). Reduction of the heart's catecholamine store, by the drug reserpine, slows or stops the heart (Bloom *et al.* 1961). The β-blocking drug sotalol, injected *in vivo*, kills or severely impairs its function.

Granulated subendothelial cells containing adrenaline occur in the *Myxine* ventricle as in some other primitive fish and Dashow and Epple (1985) suggest that in the lamprey *Petromyzon marinus* these constitute both a receptor and an effector mechanism, well placed to monitor the composition of the blood passing through the ventricle. In response to some, as yet unknown, blood-borne stimulus the cells might liberate catecholamines which exert specific effects both locally on myocytes within the heart wall and, peripherally, if liberated into the blood. Axelsson *et al.* (1990) report that injected adrenaline significantly lowers peripheral vascular resistance in *Myxine*. How, in the absence of any autonomic innervation, the myocardial stores of adrenaline in hagfish are mobilized and under what circumstances they are set free, is at present unknown. The myocardium may have, in evolution, preceded the pronephric kidney, as the primary site for catecholamine storage and release.

The heart of *Myxine* is very resistant to hypoxia. Exposure of hagfish to water with a P_{O_2} of 10.2–16.5 mm Hg for 15–35 min, a level of hypoxia which would be considered very severe for a teleost fish, leaves their cardiac output unchanged (Axelsson *et al.* 1990). We have noted the absence of a coronary circulation, and the venous blood perfusing the trabeculae must, under such circumstances, be almost totally depleted of oxygen. It seems likely that the hagfish heart can function anaerobically.

The blood

The red blood cells of *Eptatretus cirrhatus* are large: their dimensions, 19.8×13.9 μm (Wells and Forster 1989) closely match the exceptionally large cells of the roughtailed stingray (Chapter 4). The abundance of erythrocytes is low, 0.223×10^6 μl^{-1}; this value can be compared with 1.7–2.5 for the red paradise fish, 1.4–1.7 for the brown bullhead, and 2.5–3.4 for mullet (*Mugil cephalus*). The haemoglobin content of 3.0 g dl^{-1} (Table 5), is low, but not lower than that of the spiny dogfish, and the MCHC of 24.2 is as large as it is only because the blood contains rather few, but very large, red cells. The blood has an oxygen capacity of only 2.3 ml dl^{-1} (Wells *et al.* 1986), less than a quarter that of a trout (Chapter 4, Table 6). Dissolved oxygen is thus a significant component of the total blood oxygen content.

The oxygen dissociation curve is a rectangular hyperbola (Figure 18*A*) and the P_{50} is 12.3 mm Hg; a value little different from those of a number of sluggish fish such as the spiny dogfish or the trout. The P_{50} of *Myxine* blood is only 4.2 mm Hg (Bauer *et al.* 1975). In *E. cirrhatus* there is a Bohr effect, with a factor of -0.43, and this is due to CO_2 and not to H^+ sensitivity. This was demonstrated by holding the pH constant with Tris buffer and raising the P_{CO_2} from zero to 2.8 mm Hg (Figure 18*A*). A Root effect is absent and there is evidence of a mild cooperativity, with $n = 1.38$. The blood of some of the specimens of *E. stouti* studied by Bartlett (1982) had high concentrations of ADP and ATP, and a smaller amount of GTP.

Eptatretus blood has a poor buffering capacity, amounting to approximately 4 slykes, which can be compared with that of the tench, 22.5 slykes (Jensen and Weber 1982) and blue marlin, 21.3 slykes (Dobson *et al.* 1986). Part of this low value must be due to the low concentration of haemoglobin. Plasma protein is only 2.4 g%, which also must depress the buffering capacity. In addition, Ellory *et al.* (1987) have shown that the protein capnophorin is present only in very small amounts in the blood of *E. stouti.* In mammals this is known to be a major red cell transport protein involved in the exchange of Cl^- with HCO_3^-, and it is thus involved in the carriage of CO_2. The pH change in hagfish blood, in response to a challenge from KOH, was little altered in the presence of the capnophorin inhibitor H_2DIDS, compared with the large swing to alkalinity seen in the H_2DIDS-inhibited blood of the flounder.

Despite these limitations, the study of Wells *et al.* (1986) showed that the blood of *E. cirrhatus* had a significant role in transporting oxygen when the hagfish swam in a flume at 20 cm (approximately 0.4 body lengths) sec^{-1}. Dorsal aortic blood P_{O_2}, i.e. Pa_{O_2} changed little, falling only from 94.0 to 91.6 mm Hg. The ventral aortic P_{O_2}, i.e. Pv_{O_2} fell from 17.2 to 3.5 mm Hg, indicating that increased oxygen consumption in the tissues was accompanied by a greatly increased extraction of oxygen from the blood. The *E. cirrhatus* study also showed that arterial P_{CO_2} and pH did not change significantly, showing that, despite the low oxygen capacity and buffer capacity of the blood, swimming at moderate speeds is still an aerobic process.

When this species was forced to swim to exhaustion, (Davison *et al.* 1990), the blood lactate increased from 1.13 to 3.51 mmol l^{-1} and pH fell from 7.75 to 7.25. Relative to teleosts, hagfish can endure a marked acidaemia. The absence of a Root effect ensures that their modest oxygen capacity is not further reduced, as it would be in so many teleosts. Myotomal muscle lactate rose from 8.2 to 47.7 μmol g^{-1}. Hagfish can accumulate considerable amounts of lactate in their muscles when forced into activity. The white 'tongue' muscle, which is active when the hagfish is stripping food from its prey, can accumulate very high levels of lactate, and has a great intrinsic buffering capacity.

The peripheral circulation
The arterial system

The arterial system of the myxinoids is cast much along the lines of that of elasmobranch and teleost fish. There are absences associated with the lack of paired fins and the degenerate condition of the eyes: it is only in the arterial supply to the gill pouches that notable differences are to be seen. Each afferent branchial artery approaches the lateral or outer side of its gill pouch and makes a complete loop round the water duct. From this loop meridional vessels encircle the pouch, each lying above the base of one of the primary gill lamellae which project radially into the gill pouch, and on which the secondary lamellae are borne. From the outer medial surface of the gill pouch paired efferent vessels carry oxygenated blood to the common carotid artery of its side. Another unusual feature is that in the branchial region a single medial dorsal aorta exists along with the paired lateral aortae; in true fish these exist as alternatives.

The venous system

There is a curious asymmetry of the anterior cardinal veins, for the left is larger than the right; the right ductus cuvieri is absent, and the left is continuous with the sinus venosus (Figure 3A). The left posterior cardinal vein is also larger than the right. The median caudal vein is short, and divides to form the two posterior cardinals at the level of the cloaca. There is no renal portal circulation, which is perhaps related to the much reduced condition of the renal tubules. The blood supply to the nephrons is thus entirely arterial.

We have already noted, in Chapter 8, the presence of the portal heart, composed of true cardiac muscle, and located on the hepatic portal vein, just prior to its entry into the liver. The hepatic portal system also differs from that of the true fish, in that there is a vein which could enable blood to enter the portal heart from the right anterior cardinal vein. The visceral and somatic circulations in most fish are largely separate. The venous system is closely related to a second system of sinuses and ill-defined vascular spaces which surround some of the viscera.

The sinus system and the subcutaneous sinus

The myxinoids are unique in the possession of a series of blood-containing sinuses which interconnect and ultimately drain into the subcutaneous sinus. A hagfish is almost totally enveloped by a subdermal blood space which extends from the snout almost to the tip of the tail, and from the mid dorsal line to mid ventral; only in the region of the pharynx and along the line of the mucous glands is it partly absent. The layer of tissue that separates it from the water is more than just the skin, as there is some subdermal tissue as well, and it has a rich vasculature. Nevertheless, the wall is quite thin, and in *Myxine* the red blood is visible through it and can be seen to shift around within the sinus as the animal swims. The two sides communicate through the rather incomplete attachments of the sinus wall to the body along the dorsal and ventral lines.

Many smaller sinuses are present; we have already noted the hypophysio-velar sinus (Chapter 8) as the blood space above the velum, and another lies beneath the tongue. Each of the gill pouches is entirely surrounded by a branchial sinus and small tubular sinuses surround the branchial arteries and ventral aorta. Cole (1925) gives a detailed account of their interconnections, and refers to them as lymphatics. We have, in Chapter 3, noted the confusion that surrounds the use of this term in fish. They may perhaps be forerunners of the secondary blood system.

Pathways into the sinus system

Cole (1912, 1925) described curious papillae on the afferent, and particularly on the efferent arteries and dorsal aortae, which project out from the surface of the vessels, into the lumen of the surrounding sinus. Each papilla has, in its base, a small chamber which communicates with the lumen of the artery and from which several narrow canals penetrate the wall to its outer surface. They are of a diameter to hinder but not prevent the passage through them of red blood cells. When dissecting a hagfish in which the heart is still pumping, it is evident that at rest, few or no red cells enter the perivascular sinuses around the arteries and gill pouches, but the slightest pressure on these causes red cells to do so. Since Cole's two papers, these papillae have been studied by transmission EM (Elger 1987) and by scanning EM (Pohla *et al.* 1987).

These may not be the only entrances into the sinus system. Flood (1979) describes small vessels of capillary dimensions which traverse the myotomes and enter the subcutaneous sinuses. Their origins are uncertain; they may arise from the perivascular sheaths of segmental arteries.

Pathways out of the subcutaneous sinus

We have already noted, in Chapter 6, the existence of the caudal heart, which pumps blood from the caudal end of the subcutaneous sinus into the caudal vein, and the so-called 'cardinal heart', which pumps blood from the rostral end of the sinus into the anterior cardinal vein. Both of these pumps are powered by skeletal muscle. We do not know at what rate they pump but if we take the amount pumped by the single median caudal heart of a 460 g eel (P. S. Davie, personal communication), i.e. 0.5–1 ml h^{-1}, as an indication of the quantity that might be pumped by a single caudal or cardinal heart of a hagfish, and multiplied it by four, to take account of their paired nature, the total, for a 1 kg hagfish, would be 4.4–8.8 ml h^{-1}.

The haematocrit and volume of sinus blood

In 1962 Johansen *et al.* examined blood from the subcutaneous sinuses of *Myxine* and found that its haematocrit was lower than that of central venous blood, but that the plasma protein fractions were not significantly different. The subcutaneous sinus could not therefore be considered a lymphatic space. More detailed investigations of the sinus blood of *Eptatretus* show that whilst the haematocrits differ from one investigation to the next, they are always less than that of the central venous or arterial blood (Figure 50). In the central circulation the mean haematocrit was 13.5%; in the sinus it was 4.3%. Moreover, there was no significant change in sinus–blood haematocrit after 154 min of forced swimming, suggesting that the low figure was not due to red cells sedimenting within it (Forster *et al.* 1989).

When the blood volume of *E. cirrhatus* was assessed with centrally injected radio-labelled human serum albumin, ^{125}I–HSA, the label was

Figure 50. The haematocrits of subcutaneous sinus blood (○), and central blood (□) in *Eptatretus cirrhatus*. The arrow indicates the cessation of a period of swimming. Drawn from data in Forster *et al.* 1989.

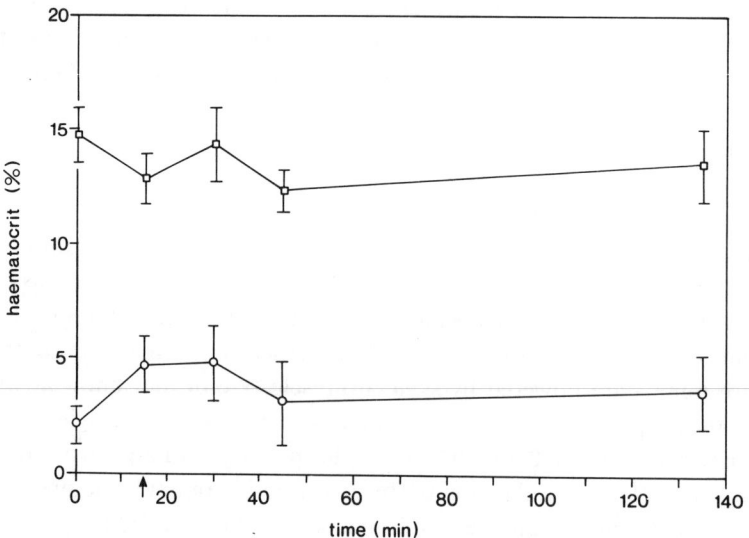

found after 2 h to be contained within a volume of 110–150 ml kg^{-1}, but this volume increased to 200 ml after 6 h and to 285 ml from 20 h onwards. These data suggest that the label escaped steadily from the blood, into the extracellular fluid, a view in accord with the report that the capillaries of hagfish are fenestrated and permeable to molecules the size of human serum albumin (Casely-Smith and Casely-Smith 1975).

The specific activity of the sinus blood increased exponentially until, after 18 h, it was the same as that in the central vessels. Moreover, the activity at the rostral and caudal ends of the sinus did not differ significantly, indicating that the entire length of the sinus, 0.8 m long, was functionally a single space.

In contrast, when the volume was assessed with ^{51}Cr labelled red cells, it did not increase markedly after 4 h; by 20–28 h the specific activity of the sinus red cells was the same as those from the central blood. From the above data it can be calculated that there was in the subcutaneous sinus of a 1 kg hagfish, 52 ml of plasma and 3.5 ml of red cells. In the central circulation there was 111 ml of plasma associated with 10.4 ml of red cells. The total blood volume was thus 177.4 ml kg^{-1}, a figure in reasonable agreement with the value of 187 ml kg^{-1}, determined for *E. stouti* by McCarthy and Conte (1966).

The investigation also brought to light the fact that the sinus is poorly circulated. The estimate of the rate of pumping blood out of the sinus, of 4.4–8.8 ml h^{-1}, set out above, might suggest that the sinus could have its blood totally replaced in 6–12 h. The data support the conclusion learned by many from practical experience, that it is certainly not an ideal site, despite its easy access, for tracer substances to be injected.

What is the role of the subcutaneous sinus?

Steffensen *et al.* (1984) presented data which suggest that in *Myxine* as much as 80% of the oxygen uptake occurs through the skin. In the earliest known stage of *Myxine* which has been studied, a shunt vessel connects the ventral with the median dorsal aorta (Fontaine 1958) and there may, at times, be advantages in bypassing the gills. Whatever the importance of cutaneous respiration may be, the blood in the subcutaneous sinus is unlikely to be there as a pool into which oxygen diffuses. The study of Hans and Tabencka (1938) shows clearly that the dermis is richly vascularized and that arterial and venous trunks enter and leave the skin at the level of the respiratory opening. There is an appreciable layer of dense connective tissue beneath the dermis and the authors give the total thickness of the wall as 322 μm. Moreover, if the sinus blood is significantly involved in transporting oxygen, the role of the papillae, as devices which hold back red cells from it, seems obscure.

A second possibility is that the subcutaneous sinus acts as a temporary dump for lactate generated during stressful exercise. The investigation of Davison *et al.* (1990) showed clearly that even following exercise to exhaustion, the levels of lactate in the sinus blood were never as great as those in the central blood, and this idea too does not seem tenable.

There is one possibility that has not been investigated. Approximately one third (32%) of the plasma is located in the sinus system, and may provide a reserve of buffering power of value after exercise. The rapid onset of a period of caudal heart pumping in *Myxine* following each bout of swimming (Figure 45), mentioned in Chapter 8 (Satchell 1984a), suggests that sinus blood has at that time a role to play in the central circulation. The low haematocrit will reduce its viscosity, a feature which may be of value in view of the low flow and the diminutive size of the pumps that empty the sinus. That of the ventral aortic blood of *E. cirrhatus* is 3.7 cp at a shear rate of 45 sec^{-1} (Wells and Forster 1989), a value closely similar to the viscosity of the blood of cold water teleosts such as the arctic char and the short and longhorn sculpin (Graham and Fletcher 1985, Graham *et al.* 1985).

Can the subcutaneous sinus be related to other subdermal vascular spaces? In Chapter 6 the secondary blood system of teleost fish was described and attention directed to the network of capillary vessels that run beneath the epidermis of their scales. In elasmobranchs the arrangement is rather different, as the placoid scales arise from deeply embedded basal plates and the capillary net lies above these and beneath the epidermis that extends between their emergent points. The lateral cutaneous vein of a fish, be it a teleost or an elasmobranch, is the longest vein in its body and in its anatomical relationships it shares five features in common with the subcutaneous sinus of a hagfish. Both lie subdermally along the flank and terminate on the tail. At this point a caudal heart pumps blood from them and returns it to the caudal vein. The caudal heart of a hagfish (Chapter 8) closely resembles that of a teleost, which is unequivocally part of the secondary blood system. In both, the caudal heart is powered by skeletal muscles driven from motor neurones in the terminal part of the spinal cord. In both the arteries by which the system makes connection with the primary arteries are equipped, at their openings, with filtering devices which tend to hold back red blood cells. As a consequence of this, in both, this blood has a lower haematocrit than that of the central circulation. The two systems may have a common origin.

Early vertebrates are believed to have been filter feeders, and the Cephalaspidomorpha were armoured, bottom-living forms found fossil in Ordovician and Silurian rocks; they are not ancestral to the myxinoids but are a parallel group. Stensiö (1932) was able to prepare ground sections of the headshield of various species and that of *Hemicyclaspis murchisoni* is figured in his monograph. It shows a tangled mesh of tubular channels which lies below the dermis; the vessels in section, are irregularly shaped and 100–200 μm in diameter. Stensiö believed them to be blood vessels, and they are connected to arteries, impressions of which occur on the underside of the headshield. They are clearly too large to be capillaries, but are some other sort of subdivided sinus-like space. Cole (1925) has described a similar sort of vascular plexus beneath the dermis of the head of *Myxine*, rostral to the auditory capsule. The evolutionary origin and function of these subdermal vascular spaces remains, at present, a mystery.

Some final speculations

A series of blood systems which included that of a segmented invertebrate such as a crayfish, a hagfish, an elasmobranch, and a teleost would demonstrate that through the course of evolution the blood has been progressively more confined to blood vessels. In many invertebrates the blood is so confined only in the median dorsal vessel and elsewhere flows in spaces between organs and muscle groups. Some of the peripheral circulation of a hagfish, such as that below the dermis of the head, is scarcely advanced from this; the numerous larger sinuses, such as the sublingual and the hypophysio-velar, are also blood spaces of this type. In an elasmobranch, the expansion of the veins to form large sinuses is still evident and is best seen close to the heart, where they are accessible to the *vis a fronte* it can generate. In most teleosts, the veins are tubular vessels, except for those that are associated with auxiliary pumping mechanisms, such as the anterior cardinal vein above the branchial pump.

We can note that the progressive reduction and enclosure of the vascular spaces is accompanied by five other changes.

(1) There is a rise in the peripheral resistance. If we calculate this, in the arbitrary units of mm Hg pressure required to propel 1 ml of blood per min per kg the value for *Myxine* is 0.75 units. Similar calculations (derived from the data in Tables 2 and 3) give values for dogfish of 3.1, and for lingcod of 3.7, units of resistance. The haemodynamic data are thus what we might expect from the observations of anatomy.

(2) There is a reduction in the blood volume. We have noted above that

that of *E. cirrhatus* is 17%, i.e. more than twice the value in elasmo-branchs, 7–8% and teleosts, 5–6%.

(3) There are changes in the heart, and particularly in the ventricle. We can note an increase in the thickness of the spongy layer, the addition of an outer compact layer, and the development of a coronary circulation. Along with this came an increase in the rate of conduction of the cardiac impulse through the ventricular myocardium. The myocardial power output increased as the ventricle became able to pump greater volumes at greater pressures.

(4) There is a rise in arterial pressure as there must be if both peripheral resistance and cardiac output increase. Table 4 (Chapter 3) brings out the point that dorsal aortic pressures in fish span a large range from the 7–12 mm Hg in hagfish, up to values in tuna comparable with those of mammals.

(5) The greater ability to control the circulatory system needs also to be noted. The autonomic nervous system of myxinoids is very poorly devel-oped; small nerve branches from the ventral spinal nerves pass to some of the vessels, although it is not known that they have a vasomotor role. The heart is without vagal or sympathetic innervation, and there are no innervated segmental arrays of chromaffin tissue. Elasmobranchs lack sympathetic innervation of the heart and gills, but sympathetic fibres innervate the chromaffin tissue of the axillary bodies and their activation can liberate catecholamines into the blood. Teleosts possess both adrenergic and cholinergic innervation of the heart and gills as well as innervated chromaffin tissue. Both elasmobranchs and teleosts can restrict visceral blood flow and thus make more blood available for active skeletal muscle. This change must have gone hand in hand with the elaboration of the gut: that of the myxinoids is a straight tube which extends the length of the animal; in the true fish it becomes more com-plexly curved and the capacity of its vascular bed must increase accordingly.

This advance of autonomic control is a feature that has accompanied, step by step, the improvement in the performance of the heart and the progress towards a completely enclosed, low-volume vessel system. Hagfish, with their extensive system of blood sinuses, have much of their blood held in vessels which are at ambient pressure. The extensive subcu-taneous sinus, separated as it is from the surrounding water only by a thin pliable covering, and containing a volume of blood many times less than its residual volume, could never be part of a vasomotor pressure-regulating system. Such a system cannot have the extent of central control from autonomic centres and a dominant branchial heart which is possible in the teleosts; simpler mechanisms have to suffice.

For such circulations one pump may not be enough. The subcutaneous sinus, holding almost a third of the blood volume at ambient pressure, could only be emptied by aspiratory pumps. We have noted the presence, fore and aft, of the cardinal heart and the caudal heart. The low systemic blood pressure that such an open system must have implies that portal systems, with their two sets of capillaries in series, will present a particular difficulty. The renal portal system of myxinoids is absent and the caudal veins run directly into the posterior cardinals. We may speculate that in the course of evolution the emergence of a renal portal system depended on the establishment of a certain minimal blood pressure. The hepatic portal circulation, we have noted, has its own unique booster pump powered by cardiac muscle.

The myxinoids undoubtedly have many primitive characteristics. We have already noted their almost total inability to osmoregulate, a feature more characteristic of invertebrates than vertebrates. Other features of hagfish anatomy must be regarded as primitive such as the absence of myelin from the nerve fibres of the peripheral and central nervous system. The presence of giant coordinating neurones in the lateral grey matter of the medulla recalls similar cells in the lancelet. Yet there are features of the heart and of the overall organization of the peripheral vessels which link them firmly to the fishes. It is their ability to bridge this gap which has attracted attention to them. Stevens and Neill (1978) remark that those fisheries biologists who have worked on tuna divide the fish, quite simply, into two groups, tunas and non-tunas. Those who have worked on the myxinoids show an equal partiality for their animals. Extremes tend to engage our attention more than the middle ground, and it is the ability of myxinoids to provide perspective to the wider view of elasmobranchs and teleosts, including the tunas, which gives them their abiding interest.

REFERENCES

Agnoletti, G., Ferrari, R., Slade, A. M., Severs, N. J. and Harris, P. (1989). Stretch-induced centrifugal movement of atrial specific granules – a preparatory step in atrial natriuretic peptide secretion. *J. mol. Cell Cardiol.* **21**, 235–9.

Ahlgren, J. A., Cheng, C.-H. C., Schrag, J. D. and DeVries, A. L. (1988). Freezing avoidance and the distribution of antifreeze glycopeptides in body fluids and tissues of Antarctic fish. *J. exp. Biol.* **137**, 549–63.

Albers, J. A. A. (1806). Über das Auge des Kabeljau *Gadus morhua*, und die Schwimmblase der Secschwalb, *Triglia hirundo*. *Götting gelehre Anz.* **2**, 681–2.

Allen, D. G. and Kentish, J. C. (1985). The cellular basis of the length–tension relation in cardiac muscle. *J. mol. Cell Cardiol.* **17**, 821–40.

Allen, W. F. (1905). The blood vascular system o the Loricati, the mail checked fishes. *Proc. Wash. Acad. Sci.* **7**, 25–157.

Allison, J. V. (1989). Blood. In *Lecture Notes on Human Physiology*, Second edition, ed. J. J. Bray, P. A. Cragg, A. D. C. Macknight, R. G. Mills and D. W. Taylor, pp. 314–70. Blackwell, Oxford.

Arbel, E. R., Liberthson, R., Langendorf, R., Pick, A., Lev, M., and Fishman, A. P. (1977). Electrophysiological and anatomical observations on the heart of the African lungfish. *Am. J. Physiol.* **232**, H24–34.

Arlock, P. (1975). Electrical activity and mechanical response in the systemic heart and the portal vein heart of *Myxine glutinosa*. *Comp. Biochem. Physiol.* **51A**, 521–2.

Ask, J. A. (1983). Comparative aspects of adrenergic receptors in the heart of lower vertebrates. *Comp. Biochem. Physiol.* **76A**, 543–52.

Audigé, J. (1910). Contribution a l'étude des reins des poissons téléostéens. *Arch. Zool. Exp. Gen.* **4**, 275–624.

Avtalion, R. R. (1981). Environmental control of the immune response in fish. *Crit. Rev. Environ. Control* **11**, 163–74.

Axelsson, M. (1988). The importance of nervous and humoral mechanisms in the control of cardiac performance in the Atlantic cod *Gadus morhua* at rest and during non-exhaustive exercise. *J. exp. Biol.* **137**, 287–303.

Axelsson, M. and Nilsson, S. (1986). Blood pressure control during exercise in the Atlantic cod, *Gadus morhua*. *J. exp. Biol.* **126**, 225–36.

Axelsson, M., Ehrenstrom, F. and Nilsson, S. (1987). Cholinergic and adrenergic influence on the teleost heart *in vivo*. *Exp. Biol.* **46**, 179–86.

Axelsson, M., Farrell, A. P. and Nilsson, S. (1990). Effect of hypoxia and drugs on the cardiovascular dynamics of the Atlantic hagfish, *Myxine glutinosa. J. exp. Biol.* (In the press.)

Bailey, J.R. and Driedzic, W. R. (1988). Perfusion-independent oxygen extraction in myoglobin-rich hearts. *J. exp. Biol.* **135**, 301–15.

Ballatori, N. and Boyer, J. L. (1988). Characteristics of L-alanine uptake in freshly isolated hepatocytes of elasmobranch *Raja erinacea. Am. J. Physiol.* **254**, R801–8.

Barnett, C. H. (1951). The structure and function of the choroidal gland of teleostean fish. *J. Anat.* **85**, 113–19.

Baroin, A., Garcia-Romeu, F., Lamarre, T. and Motais, R. (1984a). Hormone-induced co-transport with specific pharmacological properties in erythrocytes of rainbow trout, *Salmo gairdneri. J. Physiol.* **350**, 137–57.

Baroin, A., Garcia-Romeu, F., Lamarre, T. and Motais, R. (1984b). A transient sodium–hydrogen exchange system induced by catecholamines in erythrocytes of rainbow trout, *Salmo gairdneri. J. Physiol.* **356**, 21–31.

Barrett, I. and Hester, F. J. (1964). Body temperature of yellowfin and skipjack tunas in relation to sea surface temperature. *Nature, Lond.* **203**, 96–7.

Barron, M. G., Tarr, B. D. and Hayton, W. L. (1987). Temperature dependence of cardiac output and regional blood flow in rainbow trout, *Salmo gairdneri* Richardson. *J. Fish Biol.* **31**, 735–44.

Bartels, H. and Decker, B. (1985). Communicating junctions between pillar cells of the Atlantic hagfish, *Myxine glutinosa. Experientia* **41**, 1039–40.

Bartlett, G. R. (1982). Phosphates in red cells of a hagfish and a lamprey. *Comp. Biochem. Physiol.* **73A**, 141–5.

Bauer, C., Engels, U. and Paleus, S. (1975). Oxygen binding to haemoglobins of the primitive vertebrate *Myxine glutinosa* L. *Nature, London.* **256**, 66–8.

Beamish, F. W. H. (1978). Swimming capacity. In *Fish Physiology*, Vol. VII, ed. W. S. Hoar and D. J. Randall, pp. 101–87. Academic Press, New York and London.

Belaud, A. and Peyraud, C. (1971). Etude préliminaire du débit coronaire sur coeur perfusé de poisson. *J. Physiol., Paris* **63**, 165A.

Bendayan, M. (1980). Use of the protein A–gold technique for the morphological study of vascular permeability. *J. Histochem. Cytochem.* **28**, 1251–4.

Bendayan, M., Sandborn, E. B. and Rasio, E. (1974). The capillary endothelium of the rete mirabile of the swim bladder of the eel (*Anguilla anguilla*): Functional and ultrastructural aspects. *Can. J. Physiol. Pharmacol.* **52**, 613–23.

Bendayan, M., Sandborn, E. and Rasio, E. (1975). Studies of the capillary basal lamina. 1. Ultrastructure of the red body of the eel swimbladder. *Laboratory Investigation* **32**, 757–67.

Berg, T. and Steen, J. B. (1968). The mechanism of oxygen concentration in the swim bladder of the eel. *J. Physiol., London.* **195**, 631–8.

Berge, P. I. (1979). The cardiac ultrastructure of *Chimaera monstrosa* L. (Elasmo-branchii: Holocephali). *Cell Tissue Res.* **201**, 181–95.

Biörck, G. and Johansson, B. (1955). Comparative studies on temperature effects upon the electrocardiogram in some vertebrates. *Acta Physiol. Scand.* **34**, 257–72.

Birch, M. P., Carre, C. G. and Satchell, G. H. (1969). Venous return in the trunk of the Port Jackson shark, *Heterodontus portusjacksoni. J. Zool., Lond.* **159**, 31–49.

Bjenning, C., Driedzic, W. and Holmgren, S. (1989). Neuropeptide Y-like immunoreactivity in the cardiovascular nerve plexus of the elasmobranchs *Raja erinacea* and *Raja radiata. Cell Tissue Res.* **255**, 481–6.

Blaxhall, P. C. and Hood, K. (1985). Cytochemical enzyme staining of fish lymphycytes separated on a Percoll gradient. *J. Fish Biol.* **27**, 749–55.

Block, B. A. and Carey, F. G. (1985). Warm brain and eye temperatures in sharks. *J. comp. Physiol. B* **156**, 229–36.

Bloom, G., Östlund, E., Euler, U. S. von, Lishajko, F., Ritzén, M. and Adams-Ray, J. (1961). Studies on catecholamine-containing granules of specific cells in cyclostome hearts. *Acta Physiol. Scand.* **53**, suppl. 185, 1–34.

Boland, E. J. and Olson, K. R. (1979). Vascular organization of the catfish gill filament. *Cell Tissue Res.* **198**, 487–500.

Booth, J. H. (1979a). Circulation in trout gills: The relationship between branchial perfusion and the width of the lamellar blood space. *Can. J. Zool.* **57**, 2183–5.

Booth, J. H. (1979b). The effects of oxygen supply, epinephrine and acetylcholine on the distribution of blood flow in trout gills. *J. exp. Biol.* **83**, 31–9.

Boutilier, R. G., Dobson, G., Hoeger, U. and Randall, D. J. (1988). Acute exposure to graded levels of hypoxia in rainbow trout (*Salmo gairdneri*): metabolic and respiratory adaptations. *Respir. Physiol.* **71**, 69–82.

Boutilier, R. G., Iwama, G. K. and Randall, D. J. (1986). The promotion of catecholamine release in rainbow trout, *Salmo gairdneri*, by acute acidosis: interactons between red cell pH and haemoglobin oxygen-carrying capacity. *J. exp. Biol.* **123**, 145–58.

Breisch, E. A., White, F., Jones, H. M., and Laurs, R. M. (1983). Ultrastructural morphometry of the myocardium of *Thunnus alalunga. Cell Tissue Res.* **233**, 427–38.

Brett, J. R. (1965). The relation of size to the rate of oxygen consumption and sustained swimming speed of sockeye salmon (*Oncorhynchus nerka*). *J. Fish. Res. Bd Can.* **23**, 1491–501.

Brittain, T. (1987). The Root effect. *Comp. Biochem. Physiol.* **86B**, 473–81.

Brodal, A. and Fänge, R. (1963). *The Biology of* Myxine. Universitetsforlaget, Oslo.

Bundgaard, M. (1987). Tubular invaginations in cerebral endothelium and their relation to smooth-surfaced cisternae in hagfish (*Myxine glutinosa*). *Cell Tissue Res.* **249**, 359–65.

Burne, R. H. (1909). On elastic mechanisms in fishes and a snake. *Proc. zool. Soc. Lond.* 201–3.

Burne, R. H. (1923). Some peculiarities of the blood-vascular system of the porbeagle shark, *Lamna cornubica. Phil. Trans. R. Soc. London. B.* **212**, 209–59.

Butler, P. J. and Taylor, E. W. (1971). Response of the dogfish (*Scyliorhinus canicula*L.) to slowly induced and rapidly induced hypoxia. *Comp. Biochem. Physiol.* **39A**, 307–23.

Butler, P. J., Axelsson, M., Ehrenström, F., Metcalf, J. D. and Nilsson, S. (1989).

Circulating catecholamines and swimming performance in the Atlantic cod, *Gadus morhua. J. exp. Biol.* **141**, 377–87.

Butler, P. J., Metcalf, J. D. and Ginley, S. A. (1986). Plasma catecholamines in the lesser spotted dogfish and rainbow trout at rest and during different levels of exercise. *J. exp. Biol.* **123**, 409–21.

Butler, P. J., Taylor, E. W., Capra, M. F. and Davison, W. (1978). The effect of hypoxia on the levels of circulating catecholamines in the dogfish *Scyliorhinus canicula. J. comp. Physiol. B* **127**, 325–30.

Butler, P. J., Taylor, E. W. and Davison, W. (1979). The effect of long term, moderate hypoxia on acid–base balance, plasma catecholamines and possible anaerobic end products in the unrestrained dogfish *Scyliorhinus canicula. J. comp. Physiol. B* **132**, 297–303.

Butler, P. J., Taylor, E. W. and Short, S. (1977). The effect of sectioning cranial nerves V, VII, IX and X on the cardiac response of the dogfish *Scyliorhinus canicula* to environmental hypoxia. *J. exp. Biol.* **69**, 233–45.

Cameron, J. N. (1973). Oxygen dissociation and content of blood from Alaskan burbot (*Lota lota*), pike (*Esox lucius*) and grayling (*Thymallus arcticus*). *Comp. Biochem. Physiol.* **46A**, 491–6.

Cameron, J. N. (1975). Blood flow distribution as indicated by tracer microspheres in resting and hypoxic arctic grayling (*Thymallus arcticus*). *Comp. Biochem. Physiol.* **52A**, 441–4.

Cameron, J. N. (1978). Chloride shift in fish blood. *J. exp. Zool.* **206**, 289–95.

Cameron, J. N. and Davis, J. C. (1970). Gas exchange in rainbow trout (*Salmo gairdneri*) with varying blood oxygen capacity. *J. Fish. Res. Bd Can.* **27**, 1069–85.

Campbell, G. (1970). Autonomic nervous systems. In *Fish Physiology*, Vol. IV, ed. W. S. Hoar and D. J. Randall, pp. 109–32. Academic Press, New York and London.

Canelle, E. D., Campbell, G. G., Smolich, J. and Campbell, J. H. (1986). *Cardiac muscle.* Springer Verlag. Berlin.

Canty, A. A. and Farrell, A. P. (1985). Intrinsic regulation of flow in an isolated tail preparation of the ocean pout (*Macrozoarces americanus*). *Can. J. Zool.* **63**, 2013–20.

Carafoli, E. (1985). The homeostasis of calcium in heart cells. *J. mol. Cell Cardiol.* **17**, 203–12.

Carey, F. G. (1980). Warm fish. In *A Companion to Animal Physiology*, ed. C. R. Taylor, K. Johansen and L. Bolis, pp. 216–31. Cambridge University Press, Cambridge.

Carey, F. G. (1982). A brain heater in the swordfish. *Science* **216**, 1327–9.

Carey, F. G. and Lawson, K. D. (1973). Temperature regulation in free-swimming bluefin tuna. *Comp. Biochem. Physiol.* **44A**, 375–92.

Carey, F. G. and Teal, J. M. (1969a). Regulation of body temperature by the bluefin tuna. *Comp. Biochem. Physiol.* **28**, 205–13.

Carey, F. G. and Teal, J. M. (1969b). Mako and porbeagle: warm-bodied sharks. *Comp. Biochem. Physiol.* **28**, 199–204.

Carey, F. G., Kanwisher, J. W. and Stevens, E. D. (1984). Bluefin tuna warm their viscera during digestion. *J. exp. Biol.* **109**, 1–20.

Carey, F. G., Teal, J. M. and Kanwisher, J. W. (1981). The visceral temperatures of mackerel sharks (Lamnidae). *Physiol. Zool.* **54**, 334–44.

Carey, F. G., Teal, J. M., Kanwisher, J. W., Lawson, K. D. and Beckett, J. S. (1971). Warm-bodied fish. *Am. Zool.* **11**, 135–43.

Casely-Smith, J. R. and Casely-Smith, J. R. (1975). The fine structure of the blood capillaries of some endocrine glands of the hagfish, *Eptatretus stouti*: implications for the evolution of blood and lymph vessels. *Ref. suisse Zool.* **82**, 35–40.

Casely-Smith, J. R. and Mart, P. E. (1970). The relative antiquity of fenestrated blood capillaries and lymphatics, and their significance for the uptake of large molecules: an electron miroscopical investigation in an elasmobranch. *Experientia* **26**, 508–10.

Cech, J. C., Bridges, D. W., Rowell, D. M. and Balzer, P. J. (1976). Cardiovascular responses of winter flounder *Pseudopleuronectes americanus* (Walbaum), to acute temperature increase. *Can. J. Zool.* **54**, 1383–8.

Cech, J. J. Jr, Laurs, R. M. and Graham, J. B. (1984). Temperature induced changes in blood gas equilibria in the albacore, *Thunnus alalunga*, a warm-bodied tuna. *J. exp. Biol* **109**, 21–34.

Cech, J. J., Rowell, D. M. and Glasgow, J. S. (1977). Cardiovascular responses of the winter flounder *Pseudopleuronectes americanus* to hypoxia. *Comp. Biochem. Physiol.* **57A**, 123–5.

Chan, D. K. O. (1975). Cardiovascular and renal effects of urotensins I and II in the eel *Anguilla rostrata*. *Gen. comp. Endocrinol.* **27**, 52–61.

Chapman, C. B., Jensen, D. and Wildenthal, K. (1963). On circulatory control mechanisms in the Pacific hagfish. *Circulation Res.* **12**, 427–40.

Chiesa, D. F., Noseda, V. and Marchetti, R. (1962). Attivazione degli strati epicardici nel cuore de alcuni teleosti di acqua dolce. Ricerche elettrocardiografiche. *Archo. Sci. Biol.* **46**, 1–10.

Chiocchia, G. and Motais, R. (1989). Effect of catecholamines on deformability of red cells from trout: relative roles of cyclic AMP and cell volume. *J. Physiol.* **412**, 321–32.

Chow, P. H. and Chan, D. K. O. (1975). The cardiac cycle and the effects of neurohumors on myocardial contractility in the Asiatic eel, *Anguilla japonica* Timm. & Schle. *Comp. Biochem. Physiol.* **52C**, 41–5.

Claireaux, G., Thomas, S., Fievet, B. and Motais, R. (1988). Adaptive respiratory responses of trout to acute hypoxia. II. Blood oxygen carrying properties during hypoxia. *Respir. Physiol.* **74**, 91–8.

Cobb, J. L. S. (1974). Gap junctions in the heart of teleost fish. *Cell Tissue Res.* **154**, 131–4.

Cobb, J. L. S. and Santer, R. M. (1973). Electrophysiology of cardiac function in teleosts: cholinergically mediated inhibition and rebound excitation. *J. Physiol., Lond.* **230**, 561–73.

Cohen, S. L. and Kriebel, R. M. (1989). Terminal processes of serotonin neurons in the caudal spinal cord of the molly, *Poecilia latipinna*, project to the leptomeninges and urophysis. *Cell Tissue Res.* **255**, 619–25.

Cole, F. J. (1906–7). A monograph on the general morphology of the myxinoid fishes based on a study of *Myxine*. Pt II. The anatomy of the muscles. *Trans. R. Soc. Edin* **45**, 683–757.

Cole, F. J. (1912). A monograph on the general morphology of the myxinoid fishes based on a study of *Myxine*. Part IV. On some peculiarities of the afferent and efferent branchial arteries of *Myxine*. *Trans. R. Soc. Edin.* **48**, 215–30.

Cole, F. J. (1925). A monograph on the general morphology of the myxinoid fishes based on a study of *Myxine*. Part VI. The morphology of the vascular system. *Trans. R. Soc. Edin.* **54**, 309–42.

Cossins, A. R. and Kilbey, R. (1989). The seasonal modulation of Na^+/H^+ exchanger activity in trout erythrocytes. *J. exp. Biol.* **144**, 463–78.

Cossins, A. R. and Richardson, P. A. (1985). Adrenalin-induced Na^+/H^+ exchange in trout ereythrocytes and its effects upon oxygen-carrying capacity. *J. exp. Biol.* **118**, 229–46.

Crawshaw, L. I., Wollmuth, L. P. and O'Connor, C. S. (1989). Intercranial ethanol and ambient anoxia elicit selection of cooler water by goldfish. *Am. J. Physiol.* **256**, R133–7.

Crone, C. (1986). Modulation of solute permeability in microvascular endothelium. *Fed. Proc.* **45**, 77–83.

Crone, C. (1987). The Malpighi lecture. From '*Porositates carnis*' to cellular microcirculation. *Int. J. Microcirc. Clin. Exp.* **6**, 101–22.

Cumber, L. J. M., Sigel, M. M., Trauger, R. J. and Cuchens, M. A. (1982). RES structure and function of the fishes. In *The Reticuloendothelial System, a comprehensive treatise, Vol. 3. Phylogeny and ontogeny*, ed. N. Cohen and M. M. Sigel, pp. 393–422. Plenum, New York and London.

Dafré, A. L. and Danilo, W. F. (1989). Root effect hemoglobins in marine fish. *Comp. Biochem. Physiol.* **92A**, 467–71.

D'Aoust, B. G. (1970). The role of lactic acid in gas secretion in the teleost swimbladder. *Comp. Biochem. Physiol.* **32**, 637–68.

Dashow, L. and Epple, A. (1985). Plasma catecholamines in the lamprey: intrinsic cardiovascular messengers? *Comp. Biochem. Physiol.* **82C**, 119–22.

Davidson, W. S., Bartlett, S. E., Birt, T. P., Birt, V. L. and Green, J. M. (1989). Identification and purification of serum albumin from rainbow trout (*Salmo gairdneri*). *Comp. Biochem. Physiol.* **93B**, 5–9.

Davie, P. S. (1981a). Neuroanatomy and control of the caudal lymphatic heart of the short finned eel (*Anguilla australis schmidtii*). *Can. J. Zool.* **59**, 1586–92.

Davie, P. S. (1981b). Vascular resistance responses of an eel tail preparation: alpha constriction and beta dilation. *J. exp. Biol.* **90**, 65–84.

Davie, P. S. (1982). Changes in vascular and extravascular volumes of eel muscle in response to catecholamines: the function of the caudal lymphatic heart. *J. exp. Biol.* **96**, 195–208.

Davie, P. S. and Daxboeck, C. (1984). Anatomy and adrenergic pharmacology of the coronary vascular bed of Pacific blue marlin (*Makaira nigricans*). *Can. J. Zool.* **62**, 1886–8.

Davie, P. S., Forster, M. E., Davison, W. and Satchell, G. H. (1987). Cardiac function in the New Zealand hagfish, *Eptatretus cirrhatus*. *Physiol. Zool.* **60**, 233–40.

Davie, P. S., Wells, R. M. G. and Tetens, V. (1986). Effects of sustained swimming on rainbow trout muscle structure, blood oxygen transport, and lactate dehydrogenase isoenzymes: evidence for increased aerobic capacity of white muscle. *J. exp. Zool.* **237**, 159–71.

Davison, W., Baldwin, J., Davie, P. S., Forster, M. E. and Satchell, G. H. (1989). Exhausting exercise in the hagfish, *Eptatretus cirrhatus*: the anaerobic potential and the appearance of lactic acid in the blood. *Comp. Biochem. Physiol.* **95A**, 585–9.

Daxboeck, C. and Davie, P. S. (1986). Physiological investigations of marlin, pp. 50–70 in 'Fish Physiology: recent advances'. Eds. S. Nilsson and S. Holmgren. Croom Helm. London, Sydney, Dover, New Hampshire.

Daxboeck, C. and Holeton, G. F. (1978). Oxygen receptors in the rainbow trout, *Salmo gairdneri*. *Can. J. Zool.* **56**, 1254–59.

De Kock, L. and Symmons, S. (1959). A ligament in the dorsal aorta of certain fishes. *Nature, Lond.* **184**, 194.

De Roos, R. and De Roos, C. C. (1978). Elevation of plasma glucose levels by catecholamines in elasmobranch fish. *J. comp. Endocrinol.* **34**, 447–52.

De Vries, R. and De Jager, S. (1984). The gill in the spiny dogfish, *Squalus acanthias*: respiratory and nonrespiratory function. *Am. J. Anat.* **169**, 1–29.

Dobson, G. P., Wood, S. C., Daxboeck, C. and Perry, S. F. (1986). Intracellular buffering and oxygen transport in the Pacific blue marlin (*Makaira nigricans*): adaptations to high-speed swimming. *Physiol. Zool.* **59**, 150–6.

Dodd, H. and Cockett, F. B. (1976). *The Pathology and Surgery of the Veins of the Lower Limb*. Churchill Livingstone, Edinburgh, London and New York.

Doggett, T. A. and Harris, J. E. (1989). Ultrastructure of the peripheral blood leucocytes of *Oreochromis mossambicus*. *J. Fish Biol.* **33**, 747–56.

Donald, J. (1984). Adrenergic inerrvation of the gills of brown and rainbow trout. *Salmo trutta* and *S. gairdneri*. *J. Morph.* **182**, 307–16.

Donald, J. A. (1989). Vascular anatomy of the gills of the stingrays *Urolophus mucosus* and *U. paucimaculatus* (Urolophidae, Elasmobranchii). *J. Morph.* **200**, 37–46.

Dornesco, G. T. and Santa, V. (1963). La structure des aortes et des vaisseaux sanguins de la carp (*Cyprinus carpio* L.). *Anat. Anz.* **113**, 136–45.

Driedzic, W. R. (1988). Matching of cardiac oxygen delivery and fuel supply to energy demand in teleosts and cephalopods. *Can. J. Zool.* **66**, 1078–83.

Driedzic, W. R. and Stewart, J. M. (1982). Myoglobin content and the activities of enzymes of energy metabolism in red and white fish hearts. *J. comp. Physiol.* B **149**, 67–73.

Driedzic, W. R., Sidell, B. D., Stowe, D. and Branscombe, R. (1987). Matching of vertebrate cardiac energy demand to energy metabolism. *Am. J. Physiol.* **252**, R930–7.

Duff, D. W., Fitzgerald, D., Kullman, D., Lipke, D. W., Ward, J. and Olson, K. R. (1987). Blood volume and red cell space in tissues of the rainbow trout, *Salmo gairdneri*. *Comp. Biochem. Physiol.* **87A**, 393–8.

Duff, D. W. and Olson, K. R. (1986). Trout vascular and renal responses to atrial natriuretic factor and heart extracts. *Am. J. Physiol.* **251**, R639–42.

Dunel-Erb, S., Bailly, Y. and Laurent, P. (1982). Neuroepithelial cells in the fish gill primary lamellae. *J. appl. Physiol.* **53**, 1342–53.

Dunel-Erb, S., Bailly, Y. and Laurent, P. (1989). Neurons controlling the gill vasculature in five species of teleosts. *Cell Tissue Res.* **255**, 567–73.

Dunn, S. E., Murad, A. and Houston, A. H. (1989). Leucocytes and leucopoietic capacity in thermally acclimated goldfish, *Carassius auratus* L. *J. Fish Biol.* **34**, 901–11.

Eddy, F. B. (1971). Blood gas relationships in the rainbow trout *Salmo gairdneri*. *J. exp. Biol.* **55**, 695–711.

Eddy, F. B. (1973). Oxygen dissociation curves of the blood of the tench, *Tinca tinca. J. exp. Biol.* **58**, 281–93.

Eddy, F. B. (1974). Blood gases of the tench (*Tinca tinca*) in well aerated and oxygen-deficient waters. *J. exp. Biol.* **60**, 71–83.

Eddy, F. B., Lomholt, J. P., Weber, R. E. and Johansen, K. (1977). Blood respiratory properties of rainbow trout (*Salmo gairdneri*) kept in water of high CO_2 tension. *J. exp. Biol.* **67**, 37–47.

Egginton, S. and Sidell, B. D. (1989). Thermal acclimation induces adaptive changes in subcellular structure of fish skeletal muscle. *Am. J. Physiol.* **256**, R1–9.

Elger, M. (1987). The branchial circulation and the gill epithelia in the Atlantic hagfish, *Myxine glutinosa* L. *Anat. Embryol.* **175**, 489–504.

Ellis, A. E., Munroe, A. L. S. and Roberts, R. J. (1976). Defence mechanisms in fish. 1. A study of the phagocytic system and the fate of intraperitoneally injected particulate material in the plaice (*Pleuronectes platessa* L.). *J. Fish. Biol.* **8**, 67–78.

Ellory, J. C., Wolowyk, M. W. and Young, J. D. (1987). Hagfish (*Eptatretus stouti*) erythrocytes show minimal chloride transport activity. *J. exp. Biol.* **129**, 377–83.

Erskine, D. J. and Spotila, J. R. (1977). Heat-energy-budget analysis and heat transfer in the largemouth blackbass (*Micropterus salmoides*). *Physiol. Zool.* **50**, 157–69.

Eschricht, D.F. und Müller, J. (1835). Über die arteriösen und venösen Wundernetze an der Leber und einen merkwürdigen bau dieses Organes beim Thunfisch *Thyrnnus vulgaris. Abh. Dtsch. Akad. Wiss., Berlin.* pp. 1–32.

Fairbanks, M. B., Hoffert, J. R. and Fromm, P. O. (1969). The dependence of the oxygen-concentrating mechanism of the teleost eye (*Salmo gairdneri*) on the enzyme carbonic anhydrase. *J. gen. Physiol.* **54**, 203–11.

Fänge, R. (1966). Physiology of the swimbladder. *Physiol. Rev.* **46**, 299–322.

Fänge, R. (1968). The formation of eosinoplilic granulocytes in the oesophageal lymphomyeloid tissue of the elasmobranchs. *Acta zool., Stockh.* **49**, 155–61.

Fänge, R. (1976). Gas exchange in the swimbladder. In *Respiration in Amphibious Vertebrates*, ed. G. M. Hughes, pp. 189–211. Academic Press, London and New York.

Fänge, R. (1983). Gas exchange in fish swim bladder. *Rev. Physiol. Biochem. Pharmacol.* **97**, 111–58.

Fänge, R. (1984). Lymphomyeloid tissues in fishes. *Vidensk. Meddr dansk naturh. Foren.* **145**, 143–62.

Fänge, R. (1986). Lymphoid organs in sturgeons (Acipenseridae). *Vet. Immunol. Immunopathol.* **12**, 153–61.

206 *References*

Fänge, R. and Johansson-Sjöbeck, M. L. (1975). The effect of splenectomy on the haematology and on the activity of δ-aminolevulinic acid dehydratase (ALA-D) in haemopoetic tissues of the dogfish, *Scyliorhinus canicula* (Elasmobranchii). *Comp. Biochem. Physiol.* **52A**, 577–80.

Fänge, R. and Pulsford, A. (1985). The thymus of the angler fish *Lophius piscatorius* (Pisces: Teleostei) a light and electron microscopic study. In *Fish Immunology*, ed. M. J. Manning and M. F. Tatner, pp. 293–311. Academic Press, London.

Fänge, R., Holmgren, S. and Nilsson, S. (1976). Autonomic nerve control of the swimbladder of the goldsinny wrasse, *Ctenolabrus rupestris*. *Acta physiol. scand.* **97**, 292–303.

Farrell, A. P. (1979). The wind-kessel effect of the bulbus arteriosus in trout. *J. exp. Zool.* **208**, 169–74.

Farrell, A. P. (1980a). Vascular pathways in the gill of lingcod, *Ophiodon elongatus*. *Can. J. Zool.* **58**, 796–806.

Farrell, A. P. (1980b). Gill morphometrics, vessel dimensions, and vascular resistance in ling cod, *Ophiodon elongatus*. *Can. J. Zool.* **58**, 807–18.

Farrell, A. P. (1981). Cardiovascular changes in the lingcod (*Ophiodon elongatus*) following adrenergic and cholinergic drug infusions. *J. exp. Biol.* **91**, 293–305.

Farrell, A. P. (1982). Cardiovascular changes in the unanaesthetized lingcod (*Ophiodon elongatus*) during short-term, progressive hypoxia and spontaneous activity. *Can. J. Zool.* **60**, 933–41.

Farrell, A. P. (1984). A review of cardiac performance in the teleost heart: intrinsic and humoral regulation. *Can. J. Zool.* **62**, 523–36.

Farrell, A. P. (1985). A protective effect of adrenaline on the acidotic teleost heart. *J. exp. Biol.* **116**, 503–8.

Farrell, A. P. (1987). Coronary flow in a perfused rainbow trout heart. *J. exp. Biol.* **129**, 107–23.

Farrell, A. P. and Daxboeck, C. (1981). Oxygen uptake in the lingcod, *Ophiodon elongatus*, during progressive hypoxia. *Can. J. Zool.* **59**, 1272–5.

Farrell, A. P. and Graham, M. S. (1986). Effects of adrenergic drugs on the coronary circulation of Atlantic salmon (*Salmo salar*). *Can. J. Zool.* **64**, 481–4.

Farrell, A. P. and Milligan, C. L. (1986). Myocardial intracellular pH in a perfused rainbow trout heart during extracellular acidosis in the presence and absence of adrenaline. *J. exp. Biol.* **125**, 347–59.

Farrell, A. P. and Munt, B. (1983). Cholesterol levels in the blood of Atlantic salmonids. *Comp. Biochem. Physiol.* **75A**, 239–42.

Farrell, A. P. and Smith, D. J. (1981). Microvascular pressures in gill filaments of lingcod (*Ophiodon elongatus*). *J. exp. Zool.* **216**, 341–4.

Farrell, A. P. and Steffensen, J. F. (1987). Coronary ligation reduces maximum sustained swimming speed in chinook salmon *Oncorhynchus tshawytscha*. *Comp. Biochem. Physiol.* **87A**, 35–7.

Farrell, A. P., Daxboeck, C. and Randall, D. J. (1979). The effect of input pressure and flow on the pattern and resistance to flow in the isolated perfused gill of a teleost fish. *J. comp. Physiol. B* **133**, 233–40.

Farrell, A. P., Johansen, J. A. and Graham, M. S. (1988a). The role of the pericardium in cardiac performance of the trout (*Salmo gairdneri*). *Physiol. Zool.* **61**, 213–21.

Farrell, A. P., Johansen, J. A. and Saunders, R. L. (1989a). Coronary lesions in Pacific salmonids. *J. Fish Biol.* (in press).

Farrell, A. P., MacLeod, K. R. and Chancey, B. (1986b). Intrinsic mechanical properties of the perfused rainbow trout heart and the effects of catecholamines and extracellular calcium under control and acidotic conditions. *J. exp. Biol.* **125**, 319–45.

Farrell, A. P., MacLeod, K. and Driedzic, W. R. (1982). The effects of preload, after load, and epinephrine on cardiac performance in the sea raven, *Hemitripterus americanus*. *Can. J. Zool.* **60**, 3165–71.

Farrell, A. P., MacLeod, K. R., Driedzic, W. R. and Wood, S. (1983). Cardiac performance in the *in situ* perfused fish heart during extracellular acidosis: interactive effects of adrenaline. *J. exp. Biol.* **107**, 415–29.

Farrell, A. P., MacLeod, K. R. and Scott, C. (1988b). Cardiac performance of the trout (*Salmo gairdneri*) heart during acidosis: effects of low bicarbonate, lactate and cortiosol. *Comp. Biochem. Physiol.* **91A**, 271–7.

Farrell, A. P., Saunders, R. L., Freeman, H. C. and Mommsen, T. P. (1986a). Atherosclerosis in Atlantic salmon. *Atherosclerosis* **6**, 453–61.

Farrell, A. P., Small, S. and Graham, M. S. (1989b). Effect of heart rate and hypoxia on the performance of a perfused trout heart. *Can. J. Zool.* **67**, 274–80.

Farrell, A. P., Sobin, S. S., Randall, D. J. and Crosby, S. (1980). Intralamellar blood flow patterns in fish gills. *Am. J. Physiol.* **239**, R428–36.

Farrell, A. P., Wood, S., Hart, T. and Driedzic, W. R. (1985). Myocardial oxygen consumption in the sea raven, *Hemitripterus americanus*: the effects of volume loading, pressure loading and progressive hypoxia. *J. exp. Biol.* **117**, 237–50.

Favaro, G. (1906). Ricerche intorno alla morfologia ed alla sviluppo dei vasi, seni, e curori caudali nei ciclostomi e nei pesci. *Atti Ist. veneto. Sci* **65**, 1–279.

Ferguson, R. A., Tufts, B. L. and Boutilier, R. G. (1989). Energy metabolism in trout red cells: consequences of adrenergic stimulation *in vivo* and *in vitro*. *J. exp. Biol.* **143**, 133–47.

Fievet, B., Claireaux, G., Thomas, S. and Motais, R. (1988). Adaptive respiratory response of trout to acute hypoxia. III. Ion movements and pH changes in the red blood cell. *Respir. Physiol.* **74**, 99–114.

Fievet, B., Motais, R. and Thomas, S. (1987). Role of adrenergic-dependent H^+ release from red cells in acidosis induced by hypoxia in trout. *Am. J. Physiol.* **252**, R269–75.

Fletcher, G. L. (1977). Circannual cycles of blood plasma freezing point and Na^+ and Cl^- concentrations in Newfoundland winter flounder (*Pseudopleuronectes americanus*): correlation with water temperature and photoperiod. *Can. J. Zool.* **55**, 789–95.

Fletcher, G. L. and Haedrich, T. (1987). Rheological properties of rainbow trout blood. *Can. J. Zool.* **65**, 879–83.

Flood, P. R. (1979). The vascular supply of the three fibre types in the parietal trunk muscle of the Atlantic hagfish (*Myxine glutinosa* L.). *Microvascular Res.* **17**, 55–70.

Fonner, D. B., Hoffert, J. R. and Fromm, P. O. (1973). The importance of the counter current oxygen multiplier mechanism in maintaining retinal function in the teleost. *Comp. Biochem. Physiol.* **46A**, 559–67.

Fontaine, M. (1958). Classe de Cyclostomes. Formes actuelles. Super-ordres des Petromyzonoidea et des Myxinoidea. In Tome XIII *Traité de Zoologie, anatomie, systématique, biologie*, Tome XIII, ed. P. P. Grassé, pp. 13–172. Masson et Cie, Paris.

Forster, M. E. (1989). Performance of the heart of the hagfish *Eptatretus cirrhatus*. *Fish. Physiol. Biochem.* **6**, 327–31.

Forster, M. E., Davison, W., Satchell, G. H. and Taylor, H. H. (1989). The subcutaneous sinus of the hagfish, *Eptatretus cirrhatus* and its relation to the central circulating blood volume. *Comp. Biochem. Physiol.* **93A**, 607–12.

Forster, M. E., Davie, P. S., Davison, W., Satchell, G. H. and Wells, R. M. G. (1988). Blood pressures and heart rates in swimming hagfish. *Comp. Biochem. Physiol.* **89A**, 247–50.

Forster, R. E. and Steen, J. B. (1969). The rate of the 'Root shift' in eel red cells and eel haemoglobin solutions. *J. Physiol., Lond.* **204**, 259–82.

Franklin, K. J. (1937). *A Monograph on Veins.* Charles C. Thomas, Springfield.

Fuchs, D. A. and Albers, C. (1988). Effect of adrenaline and blood gas conditions on red cell volume and intra-erythrocytic electrolytes in the carp, *Cyprinus carpio*. *J. exp. Biol.* **137**, 457–77.

George, C. J., Ellis, A. E. and Bruno, D. W. (1982). On remembrance of the abdominal pores in the rainbow trout, *Salmo gairdneri* Richardson and some other salmonid spp. *J. Fish Biol.* **21**, 643–7.

Georgi, T. A. and Beedle, D. (1978). The histology of the excretory kidney of the paddlefish, *Polyodon spathula*. *J. Fish Biol.* **13**, 587–90.

Gesser, H. (1977). The effects of hypoxia and reoxygenation on force development in myocardia of carp and rainbow trout: protective effects of CO_2/HCO_3. *J. exp. Biol.* **69**, 199–206.

Gesser, H. and Poupa, O. (1973). The lactate dehydrogenase system in the heart and skeletal muscle of fish: a comparative study. *Comp. Biochem. Physiol.* **46B**, 683–90.

Gesser, H. and Poupa, O. (1974). Relations between heart muscle enzyme pattern and directly measured tolerance to acute anoxia. *Comp. Biochem. Physiol.* **48A**, 97–103.

Gingerich, W. H., Pityer, R. A. and Rach, J. J. (1987). Estimates of plasma, packed cell and total blood volume in tissues of the rainbow trout (*Salmo gairdneri*). *Comp. Biochem. Physiol.* **87A**, 251–6.

Giles, M. A. and Randall, D. J. (1980). Oxygenation characteristics of the polymorphic hemoglobins of coho salmon (*Oncorhynchus kisutch*) at different developmental stages. *Comp. Biochem. Physiol.* **65A**, 265–71.

Goodrich, E. S. (1930). Studies on the Structure and Development of Vertebrates. Macmillan & Co., London.

Graham, M. and Farrell, A. P. (1989). The effect of temperature acclimation and adrenaline on the performance of a perfused trout heart. *Physiol. Zool.* **62**, 38–61.

Graham, M. S. and Fletcher, G. L. (1983). Blood and plasma viscosity of winter flounder: influence of temperature, red cell concentration, and shear rate. *Can. J. Zool.* **61**, 2344–50.

Graham, M. S. and Fletcher, G. L. (1985). On the low viscosity blood of two cold water, marine sculpins. *J. comp. Physiol. B* **155**, 455–9.

Graham, M. S., Fletcher, G. L. and Haedrich, R. L. (1985). Blood viscosity in Arctic fishes. *J. exp. Zool.* **234**, 157–60.

Greaney, G. S. and Powers, D. A. (1977). Cellular regulation of an allosteric modifier of fish haemoglobin. *Nature, Lond.* **270**, 73–4.

Grodzinski, Z. (1933). Bemerkungen über das Lymphgefässsystem der Myxine glutinosa. *Bull. Acad. Pol. Sci. Let. B* 221–36.

Hans, M. and Tabencka, Z. (1938). Über die Blutgefässe der Haut von Myxine glutinosa. *Bull. Acad. Pol. Sci. Lett. B* **69–77**.

Hardisty, M. W. (1982). Lampreys and hagfish; analysis of cyclostome relationships. In *The Biology of Lampreys*, Vol. 4B, ed. M. W. Hardisty and I. C. Potter, pp. 165–259. Academic Press, London and New York.

Hargens, A. R., Millard, R. W. and Johansen, K. (1974). High capillary permeability in fishes. *Comp. Biochem. Physiol.* **48A**, 675–80.

Harrington, J. P. (1986). Structural and functional studies of the king salmon, *Oncorhynchus tshawytscha* hemoglobins. *Comp. Biochem. Physiol.* **84B**, 111–16.

Hart, S., Wrathmell, A. B. and Harris, J. E. (1986). Ontogeny of gut-associated lymphoid tissue (GALT) in the dogfish *Scyliorhinus canicula* L. *Vet. Immunol. Immunopathol.* **12**, 107–16.

Hartman, F. A. and Lessler, M. A. (1964). Erythrocyte measurements in fishes, amphibia and reptiles. *Biol. Bull. mar. biol. Lab., Woods Hole* **126**, 83–8.

Harvey, W. (1649). *De circulatione sanguinis*. Another exercitation to John Riolan. In *The anatomical exercises of Dr William Harvey*, ed. G. Keynes, pp. 145–93. Nonsuch Press, London.

Haywood, G. P., Isaia, J. and Maetz, J. (1977). Epinephrine effects on branchial water and urea flux in rainbow trout. *Am. J. Physiol.* **232**, R110–15.

Heath, A. G. and Hughes, G. M. (1973). Cardiovascular and respiratory changes during heat stress in rainbow trout (*Salmo gairdneri*). *J. exp. Biol.* **59**, 323–38.

Helle, K. B. and Storesund, A. (1975). Ultrastructural evidence for a direct connection between the myocardial granules and the sarcoplasmic reticulum in the cardiac ventricle of *Myxine glutinosa* (L.). *Cell Tissue Res.* **163**, 353–63.

Helle, K. B., Miralto, A., Pihl, K. E. and Tota, B. (1983). Structural organization of the normal and anoxic heart of *Scyllium stellare*. *Cell Tissue Res.* **231**, 399–414.

Henry, R. P., Smatresk, N. J. and Cameron, J. N. (1988). The distribution of branchial carbonic anhydrase and the effects of gill and erythrocyte carbonic anhydrase inhibition in the channel catfish *Ictalurus punctuatus*. *J. exp. Biol.* **134**, 201–18.

Herraez, M. P. and Zapata, A. G. (1986). Structure and function of the melano-macrophage centres of the goldfish *Carassius auratus*. *Vet. Immunol. Immunopathol.* **12**, 117–26.

Hicks, J. W. and Badeer, H. S. (1989). Siphon mechanism in collapsible tubes: application to circulation of the giraffe head. *Am. J. Physiol.* **256**, R567–71.

Hill, A. V. (1910). The possible effects of the aggregation of the molecules of haemoglobin on its dissociation curves. *J. Physiol., Lond.* **40**, iv–vii.

Hine, P. M. and Wain, J. M. (1988a). Observations on the granulocyte peroxidase of teleosts: a phylogenetic perspective. *J. Fish Biol.* **33**, 247–54.

Hine, P. M. and Wain, J. M. (1988b). Characterization of inflammatory neutrophils induced by bacterial endotoxin in the blood of eels, *Anguilla australis. J. Fish Biol.* **32**, 579–92.

Hine, P. M. and Wain, J. M. (1989). Observations on eosinophilic granule cells in peritoneal exudates of eels, *Anguilla australis. J. Fish Biol.* **34**, 841–53.

Hipkins, S. F. (1985). Adrenergic responses of the cardiovascular system of the eel *Anguilla australis*, in vivo. *J. exp. Zool.* **235**, 7–20.

Hoffert, J. R. and Fromm, P. O. (1972). Teleost retinal metabolism as affected by acetazolamide (36298). *Proc. Soc. exp. Biol. Med.* **139**, 1060–4.

Hoffert, J. R. and Fromm, P. O. (1973). Effect of acetazolamide on some hematological parameters and ocular oxygen concentration in rainbow trout. *Comp. Biochem. Physiol.* **45A**, 371–78.

Hoffman, E. A. and Ritman, E. L. (1985). Invariant total heart volume in the intact thorax. *Am. J. Physiol.* **249**, H883–90.

Holbert, P. W., Boland, E. J. and Olson, K. R. (1979). The effect of epinephrine and acetylcholine on the distribution of red cells within the gills of the channel catfish (*Ictalurus punctatus*). *J. exp. Biol.* **79**, 135–46.

Holmgren, S. (1977). Regulation of the heart of a teleost, *Gadus morhua*, by autonomic nerves and circulating catecholamines. *Acta physiol. scand.* **99**, 62–74.

Holmgren, S. and Nilsson, S. (1976). Effects of denervation, 6-hydroxydopamine and reserpine on the cholinergic and adrenergic responses of the spleen of the cod, *Gadus morhua. Eur. J. Pharmacol.* **39**, 53–9.

Holeton, G. F. (1977). Constancy of arterial blood pH during CO-induced hypoxia in the rainbow trout. *Can. J. Zool.* **55**, 1010–13.

Holeton, G. F. and Randall, D. J. (1967). The effect of hypoxia on the partial pressure of gases in the blood and water afferent and efferent to the gills in trout. *J. exp. Biol.* **46**, 317–27.

Huang, T. F. (1973). The action potential of the myocardial cells of the golden carp. *Jap. J. Physiol.* **23**, 529–40.

Hughes, G. M. (1966). The dimensions of fish gills in relation to their function. *J. exp. Biol.* **45**, 177–95.

Hughes, G. M. and Kikuchi, Y. (1984). Effects of *in vivo* and *in vitro* changes in P_{O_2} on the deformability of red cells of rainbow trout (*Salmo gairdneri* R.). *J. exp. Biol.* **111**, 253–7.

Hughes, G. M. and Morgan, M. (1973). The structure of fish gills in relation to their respiratory function. *Biol. Rev., Cambridge* **48**, 419–75.

Hughes, G. M., Kikuchi, Y. and Watari, H. (1982). A study of the deformability of red blood cells of a teleost fish, the yellowtail (*Seriola quinqueradiata*), and a comparison with human erythrocytes. *J. exp. Biol.* **96**, 209–20.

Hunt, T. C. Rowley, A. F. (1986). Studies on the reticulo-endothelial system of the dogfish, *Scyliorhinus canicula*. Endocytic activity of fixed cells in the gills and peripheral blood leucocytes. *Cell Tissue Res.* **244**, 215–26.

Hyrtl, J. (1843). Über die Caudal und Kopf-Sinuse der Fische, und das damit zusammenhängende Seitengefass-System. *Müller's Arch. Anat. Physiol.* **10**, 224–40.

Ingermann, R. L. and Terwilliger, R. C. (1982). Presence and possible function of Root effect hemoglobins in fishes lacking functional swimbladders. *J. exp. Zool.* **220**, 171–7.

Isaia, J., Payan, P. and Girard, J. P. (1979). A study of the water permeability of the gills of freshwater- and seawater-adapted trout (*Salmo gairdneri*): mode of action of epinephrine. *Physiol. Zool.* **52**, 269–79.

Ishimatsu, A., Iwama, G. K. and Heisler, N. (1988). *In vivo* analysis of partitioning of cardiac output between systemic and central venous sinus circuits in rainbow trout: a new approach using chronic cannulation of the branchial vein. *J. exp. Biol.* **137**, 75–88.

Janssens, P. A. and Waterman, J. (1988). Hormonal regulation of gluconeogenesis and glycogenolysis in carp (*Cyprinus carpio*) liver pieces cultured *in vivo*. *Comp. Biochem. Physiol.* **91A**, 451–5.

Jasinski, A. and Kilarski, W. (1971). Capillaries in the rete mirabile and in the gas gland of the swimbladder in fishes, *Perca fluviatilis* L. and *Misgurnus fossilis* L. An electron microscopic study. *Acta Anat.* **78**, 210–23.

Jensen, D. (1961). Cardioregulation in an aneural heart. *Comp. Biochem. Physiol.* **2**, 181–201.

Jensen, D. (1965). The aneural heart of the hagfish. *Ann. N.Y. Acad. Sci.* **127**, 443–58.

Jensen, F. B. (1989). Hydrogen ion equilibria in fish haemoglobins. *J. exp. Biol.* **143**, 225–34.

Jensen, F. B. and Weber, R. E. (1982). Respiratory properties of tench blood and haemoglobin. Adaptation to hypoxic–hypercapnic water. *Mol. Physiol.* **2**, 235–50.

Johansen, K. (1962). Cardiac output and pulsatile aortic flow in the teleost, *Gadus morhua. Comp. Biochem. Physiol.* **7**, 169–74.

Johansen, K. (1963). The cardiovascular system of *Myxine glutinosa* L. In *The biology of* Myxine, ed. A. Brodal and R. Fänge, pp. 289–316. Universitetsforlaget, Oslo.

Johansen, K. and Hanson, D. (1967). Hepatic vein sphincters in elasmobranchs and their significance in controlling hepatic blood flow. J. exp. Biol. **46**, 195–203.

Johansen, K. and Pettersson, K. (1981). Gill O_2 consumption in a teleost fish, *Gadus morhua. Respir. Physiol.* **44**, 277–84.

Johansen, K. and Strahan, R. (1963). The respiratory system of *Myxine glutinosa* L. In *The biology of* Myxine, ed. A. Brodal and R. Fänge, pp. 352–71. Universitetsforlaget, Oslo.

Johansen, K., Fänge, R. and Johannessen, M. W. (1962). Relations between blood, sinus fluid and lymph in *Myxine glutinosa* L. *Comp. Biochem. Physiol.* **7**, 23–8.

Johansen, K., Franklin, D. L. and Van Citters, R. L. (1966). Aortic blood flow in free-swimming elasmobranchs. *Comp. Biochem. Physiol.* **19**, 151–60.

Johansen, K., Lykkeboe, G., Weber, R. E. and Maloiy, G. M. O. (1976). Respiratory properties of blood in awake and estivating lungfish, *Protopterus amphibius. Respir. Physiol.* **27**, 335–45.

Johansen, K., Maloiy, G.M.O. and Lykkeboe, G. (1975). A fish in extreme alkalinity. *Respir. Physiol.* **24**, 159–62.

Johnston, I. A. (1982). Capillarisation, oxygen diffusion distances and mitochondrial content of carp muscles following acclimation to summer and winter temperatures. *Cell. Tissue Res.* **222**, 325–37.

Johnston, I. A. and Goldspink, G. (1973). Quantitative studies on muscle glycogen utilisation during sustained swimming in crucian carp (*Carassius carassius* L.). *J. exp. Biol.* **59**, 607–15.

Jones, D. R., Langille, B. L., Randall, D. J. and Shelton, G. (1974). Blood flow in dorsal and ventral aortas of the cod, *Gadus morhua. Am. J. Physiol.* **226**, 90–5.

Jones, D. R., Brill, R. W. and Mense, D. C. (1986). The influence of blood gas properties on gas tensions and pH of ventral and dorsal aortic blood in free-swimming tuna, *Euthynnus affinis. J. exp. Biol.* **120**, 201–13.

Kampmeier, O. F. (1969). *Evolution and Comparative Morphology of the Lymphatic System.* Charles C. Thomas, Springfield.

Kerstens, A., Lomholt, J. P. and Johansen, K. (1979). The ventilation, extraction and uptake of oxygen in undisturbed flounders, *Platichthys flesus*: responses to hypoxia acclimation. *J. exp. Biol.* **83**, 169–79.

Kiceniuk, J. W. and Jones, D. R. (1977). The oxygen transport system in trout (*Salmo gairdneri*) during sustained exercise. *J. exp. Biol.* **69**, 247–60.

Kisch, B. (1948). Electrographic investigations of the heart of fish. *Exp. Med. Surg.* **6**, 31–62.

Kishinouye, K. (1923). Contributions to the comparative study of the so-called scombroid fishes. *Univ. Facul. Agric. Bull., Tokyo* **8**, 283–475.

Klawe, W. L., Barrett, I. and Klawe, B. M. H. (1963). Haemoglobin content of the blood of six species of scombroid fishes. *Nature, Lond.* **198**, 96.

Kloas, W., Flügge, G., Fuchs, E. and Stolle, H. (1988). Binding sites for atrial natriuretic peptide in the kidney and aorta of the hagfish (*Myxine gutinosa*). *Comp. Biochem. Physiol.* **91A**, 685–8.

Kono, M. and Hashimoto, K. (1977). Organic phosphates in the erythrocytes of fishes. *Bull. Jap. Soc. Sci. Fish.* **43**, 1307–12.

Kryvi, H., Flood, P. R. and Gulyaev, D. (1980). The ultrastructure and vascular supply of the different fibre types in the axial muscle of the sturgeon, *Acipenser stellatus* Pallas. *Cell Tissue Res.* **212**, 117–26.

Laffont, J. and Labat, R. (1966). Action de l'adrénaline sur la fréquence cardiaque de la carpe commune. Effect de la température du milieu sur l'intensité de la réaction. *J. Physiol., Paris* **58**, 351–5.

Lagerstrand, G., Mattisson, A. and Poupa, O. (1983). Studies on the calcium paradox phenomenon in cardiac muscle strips of poikolotherms. *Comp. Biochem. Physiol.* **76A**, 601–13.

Lai, N. G., Graham, J. B., Lowell, W. R. and Laurs, R. M. (1987). Pericardial and vascular pressures and blood flow in the albacore tuna (*Thunnus alalunga*). *Exp. Biol., Berlin* **46**, 187–92.

Lamers, C. H. J. (1986). Histophysiology of a primary immune response against *Aeromonas hydrophila* in carp (*Cyprinus carpio* L.). *J. exp. Zool.* **238**, 71–80.

Lanctin, H. P., McMorran, L. E. and Driedzic, W. R. (1980). Rates of glucose and lactate oxidation by the perfused isolated trout (*Salvelinus fontinalis*) heart. *Can. J. Zool.* **58**, 1708–11.

Lansing, A. I. (1959). Elastic tissue. In *The arterial wall*, ed. A. I. Lansing. Wilkins and Wilkins, Baltimore.

Larsson, A. and Fänge, R. (1977). Cholesterol and free fatty acids (FFA) in the blood of marine fish. *Comp. Biochem. Physiol.* **57B**, 191–6.

Larsson, A., Johansson-Sjöbeck, M.-L. and Fänge, R. (1976). Comparative study of some haematological and biochemical blood parameters in fishes from the Skagerrak. *J. Fish Biol.* **9**, 425–40.

Laurent, P. (1962). Contribution à l'étude morphologique et physiologique de l'innervation du coeur des téléostéens. *Arch. Anat. micros. Morph. exp.* **51**, 337–458.

Laurent, P. (1967). Le pseudobranchie des téléostéens: prevues electrophysiologique de ses fonctions, chémoréceptrice et baroréceptrice. *C. R. Acad. Sci. Paris* **264**, 1879–92.

Laurent, P. (1974). Pseudobranchial receptors in teleosts. In *Electroreceptors and Other Specialised Receptors in Lower Vertebrates, Vol III/3, Handbook of sensory physiology*, ed. A. Fessard, pp. 279–96. Springer-Verlag, Berlin, Heidelberg and New York.

Laurent, P. and Dunel, S. (1966). Recherches sur l'innervation de la pseudobranchie des téléostéens. *Arch. Anat. microsc. Morph. exp.* **55**, 634–56.

Laurent, P. and Dunel, S. (1976). Functional organization of the teleost gill. I. Blood pathways. *Acta zool., Stockh.* **57**, 189–209.

Laurent, P. and Dunel, S. (1980). Morphology of gill epithelia in fish. *Am. J. Physiol.* **238**, R147–59.

Leak, L. V. (1969). Electron microscopy of cardiac tissue in a primitive vertebrate *Myxine glutinosa*. *J. Morph.* **128**, 131–58.

Lederis, K. (1984). The fish urotensins: hypophyseal and peripheral actions in fishes and mammals. *Frontiers in Neuroendocrinology* **8**, 247–63.

Leknes, I. L. (1980). Ultrastructure of atrial endocardium and myocardium in three species of gadidae (Teleostei). *Cell Tissue Res.* **210**, 1–10.

Leknes, I. L. (1981). Ultrahistochemical studies on the moderately electron dense bodies in teleostean endocardial cells. *Histochemistry* **72**, 211–14.

Lemanski, L. F., Fitts, E. P. and Marx, B. S. (1975). Fine structure of the heart in the Japanese medaka, *Oryzias latipes*. *J. Ultrastruct. Res.* **53**, 37–65.

Lewis, J. H. (1972). Comparative hemostasis: studies on elasmobranchs. *Comp. Biochem. Physiol.* **42A**, 233–40.

Licht, J. H. and Harris, W. S. (1973). The structure, composition and elastic properties of the teleost bulbus arteriosus in the carp *Cyprinus carpio*. *Comp. Biochem. Physiol.* **46A**, 699–708.

Ling, N. and Wells, R. M. G. (1985). Plasma catecholamines and erythrocyte swelling following capture stress in a marine teleost fish. *Comp. Biochem. Physiol.* **82C**, 231–4.

Lobb, C. J. (1986). Structural diversity of channel catfish immunoglobulins. *Vet. Immunol. Immunopathol.* **12**, 7–12.

Loretz, C. A. and Bern, H. A. (1980). Ion transport by the urinary bladder of the gobiid teleost, *Gillichthys mirabilis*. *Am. J. Physiol.* **239**, R415–23.

Loretz, C. A., Freel, R. W. and Bern, H. A. (1983). Specificity of responses of intestinal ion transport systems to a pair of natural peptide hormone analogs: somatostatin and urotensin II. *Gen. comp. Endocrinol.* **52**, 198–206.

Lutz, B. R. and Wyman, L. C. (1932). Reflex cardiac inhibition of branchiovascular origin in the elasmobranch *Squalus acanthias*. *Biol. Bull. mar. biol. Lab., Woods Hole* **62**, 10–16.

Lykkeboe, G. and Weber, R. E. (1978). Changes in the respiratory properties of the blood of carp induced by diurnal variation in ambient oxygen tension. *J. comp. Physiol. B* **128**, 117–25.

MacArthur, J. I., Fletcher, T. C. and Thomson, A. W. (1983). Distribution of radiolabelled erythrocytes and the effect of temperature on clearance in the plaice (*Pleuronectes platessa* L.). *J. Reticuloendothelial Soc.* **34**, 13–21.

McCarthy, J. E. and Conte, F. P. (1966). Determination of the volume of the vascular and extravascular fluid in the Pacific hagfish, *Eptatretus stouti* (Lockington). *Am. Zool.* **6**, 605.

Mainoya, J. R. and Bern, H. A. (1982). Effects of teleost urotensins on intestinal absorption of water and NaCl in tilapia, *Sarotherodon mossambicus*, adapted to fresh water or sea water. *Gen. comp. Endocrinol.* **47**, 54–8.

Mallatt, J. (1984). Early vertebrate evolution: pharyngeal structure and the origin of gnathostomes. *J. Zool., Lond.* **204**, 169–83.

Maren, T. H. (1967). Special body fluids of the elasmobranch. In *Sharks, Skates and Rays*, eds. P. W. Gilbert, R. F. Mathewson and D. P. Rall, pp. 287–92. Johns Hopkins Press, Baltimore.

Matthiessen, P. and Brafield, A. A. (1977). Uptake and loss of dissolved zinc by the stickleback *Gasterosteus aculeatus* L. *J. Fish Biol.* **10**, 399–410.

Metcalf, J. D. and Butler, P. J. (1984). On the nervous regulation of gill blood flow in the dogfish (*Scyliorhinus canicula*). *J. exp. Biol.* **113**, 253–67.

Metcalf, J. D. and Butler, P. J. (1988). The effects of alpha- and beta-adrenergic blockade on gas exchange in the dogfish (*Scyliorhinus canicula* L.) during normoxia and hypoxia. *J. comp. Physiol. B* **158**, 39–44.

Metcalf, J. D. and Butler, P. J. (1989). The use of alpha-methyl-*p*-tyrosine to control circulating catecholamines in the dogfish (*Scyliorhinus canicula*): the effects on gas exchange in normoxia and hypoxia. *J. exp. Biol.* **141**, 21–32.

Michel, C. C. (1983). The effects of certain proteins on capillary permeability to fluid and macromolecules. In *Pathogenicity of Cationic Proteins*, ed. P. P. Lambert, P. Bergmann and R. Beauwens, pp. 125–40. Raven Press, New York.

Michel, C. C. (1988). Capillary permeability and how it may change. *J. Physiol.* **404**, 1–29.

Midtun, B. (1980). Ultrastructure of atrial and ventricular myocardium in the pike *Esox lucius* L. and mackerel *Scomber scombrus* L. (Pisces). *Cell Tissue Res.* **211**, 41–50.

Milligan, C. L. and Wood, C. M. (1987). Regulation of blood oxygen transport and red cell pH$_i$ after exhaustive activity in rainbow trout (*Salmo gairdneri*) and starry flounder (*Platichthys stellatus*). *J. exp. Biol.* **133**, 263–82.

Mislin, H. (1969). Zur Funktionsanalyse des lymphatischen Kaudalherzens beim Aal (*Anguilla anguilla* L.). *Rev. suisse Zool.* **67**, 262–9.

Mosse, P. R. L. (1978). The distribution of capillaries in the somatic musculature of two vertebrate types with particular reference to teleost fish. *Cell Tissue Res.* **187**, 281–303.

Mott, J. C. (1951). Some factors affecting the blood circulation in the common eel (*Anguilla anguilla*). *J. Physiol., Lond.* **114**, 387–98.

Mulcahy, M. F. (1970). Blood values in the pike *Esox lucius* L. *J. Fish Biol.* **2**, 203–9.

Myhre, K. and Steen, J. B. (1977). The effect of plasma proteins on the capillary permeability in the *rete mirabile* of the eel (*Anguilla vulgaris* L.). *Acta physiol. scand.* **99**, 98–104.

Myklebust, R. and Kryvi, H. (1979). Ultrastructure of the heart of the sturgeon *Acipenser stellatus* (Chondrostei). *Cell Tissue Res.* **202**, 431–8.

Nakatani, R. E. (1966). Biological responses of the rainbow trout (*Salmo gairdneri*) ingesting [65]zinc. In *Disposal of Radioactive Wastes in Seas, Oceans and Surface Waters*, pp. 809–23. International Atomic Energy Agency, Vienna.

Nekvasil, N. P. and Olson, K. R. (1986). Plasma clearance, metabolism, and tissue accumulation of ^{3}H-labelled catecholamines in trout. *Am. J. Physiol.* **250**, R519–25.

Neuville, H. (1901). A l'étude de la vascularisation intestinale chez les cyclostomes et les sélaciens. *Ann. Sci. Naturelles, Zool.* **13**, 1–116.

Nichols, W. W., Mehta, J. L., Donnelly, W. D., Lawson, D., Thompson, L. and ter Riet, M. (1988). Reduction in coronary vasodilator reserve following coronary occlusion and reperfusion in anesthetized dog: role of endothelium-derived relaxing factor, myocardial neutrophil infiltration and prostaglandins. *J. mol. Cell. Cardiol.* **20**, 943–54.

Nikinmaa, M. (1983). Adrenergic regulation of haemoglobin oxygen affinity in rainbow trout red cells. *J. comp. Physiol. B.* **152**, 67–72.

Nikinmaa, M. and Huestis, W. H. (1984). Adrenergic swelling of nucleated erythrocytes: cellular mechanisms in a bird, domestic goose, and two teleosts, striped bass and rainbow trout. *J. exp. Biol.* **113**, 215–24.

Nilsson, S. (1983). *Autonomic Nerve Function in the Vertebrates*. Springer Verlag, Berlin and New York.

Nilsson, S. (1984). Adrenergic control systems in fish. *Mar. Biol. Lett.* **5**, 127–46.

Nilsson, S. (1972). Autonomic vasomotor innervation in the gas gland of the swimbladder of a teleost (*Gadus morhua*). *Comp. gen. Pharmacol.* **3**, 371–5.

Nilsson, S. and Fänge, R. (1967). Adrenergic receptors in the swimbladder and gut of a teleost (*Anguilla anguilla*). *Comp. Biochem. Physiol.* **23**, 661–4.

Nilsson, S. and Grove, D. J. (1974). Adrenergic and cholinergic innervation of the spleen of the cod: *Gadus morhua*. *Eur. J. Pharmacol.* **28**, 135–43.

Nilsson, S. and Pettersson, K. (1981). Sympathetic nervous control of blood flow in the gills of the Atlantic cod, *Gadus morhua*. *J. comp. Physiol.* **144**, 157–63.

Nilsson, S., Abrahamsson, T. and Grove, D. J. (1976). Sympathetic nervous control of adrenaline release from the head kidney of the cod, *Gadus morhua*. *comp. Biochem. Physiol.* **55C**, 123–7.

Noble, R. W., Pennelly, R. R. and Riggs, A. (1975). Studies on the functional properties of the hemoglobin from the benthic fish, *Antimora rostrata*. *J. Comp. Physiol.* **52B**, 75–81.

Norsk, P., Foldager, N., Petersen, F. B., Larsen, B. E. and Johansen, T. S. (1987). Central venous pressure in humans during short periods of weightlessness. *J. appl. Physiol.* **63**, 2433–7.

Ogilvy, C. S. and DuBois, A. B. (1982). Effect of tilting on blood pressure and interstitial fluid pressures of bluefish and smooth dogfish. *Am. J. Physiol.* **242**, R70–6.

Olesen, S. P., Clapham, D. E. and Davies, P. F. (1988). Haemodynamic shear stress activates a K^+ current in vascular endothelial cells. *Nature, Lond.* **331**, 168–70.

Olson, K. R. (1981). Morphology and vascular anatomy of the gills of a primitive air-breathing fish, the bowfin (*Amia calva*). *Cell Tissue Res.* **218**, 499–517.

Olson, K. R. (1983). Effects of perfusion pressure on the morphology of the central sinus in the trout gill filament. *Cell Tissue Res.* **232**, 319–26.

Olson, K. R. (1984). Distribution of flow and plasma skimming in isolated perfused gills of three teleosts. *J. exp. Biol.* **109**, 97–108.

Olson, K. R. and Kent, B. (1980). The microvasculature of the elasmobranch gill. *Cell Tissue Res.* **209**, 49–63.

Olson, K. R. and Meisheri, K. D. (1989). Effects of atrial natriuretic factor on isolated arteries and perfused organs of trout. *Am. J. Physiol.* **256**, R10–18.

Ostádal, B. und Schiebler, T. H. (1971). Über die terminale Strombahn in Fischherzen. *Z. Anat. Entwickl. Gesch.* **134**, 101–10.

Owman, C. H. and Rudeberg, C. (1970). Light fluorescence and electron microscopic studies on the pineal organ of the pike *Esox lucius* L. with special regard to 5-hydroxytryptamine. *Z. Zellforsch* **107**, 522–50.

Paintal, A. S. (1955). Impulses in vagal afferent fibres from specific pulmonary deflation receptors. The response of these receptors to phenyl-di-guanide, potato starch, 5-hydroxytryptamine and nicotine and their role in respiratory and cardiovascular reflexes. *Q. J. exp. Physiol.* **40**, 89–111.

Paintal, A. S. (1957). The location and excitation of pulmonary deflation receptors by chemical substances. *Q. J. exp. Physiol.* **42**, 56–71.

Paintal, A. S. (1969). Mechanisms of stimulation of Type J pulmonary receptors. *J. Physiol., Lond.* **203**, 511–32.

Paintal, A. S. (1970). The mechanism of excitation of Type J receptors and the J reflex. In *Breathing, Ciba Foundation Hering–Breuer centenary symposium*, ed. R. Porter, pp. 59–76. Churchill, London.

Pankhurst, L. J. and Goss, D. J. (1984). Ligand binding kinetic studies on the hybrid haemoglobin alpha (human) beta (carp). A haemoglobin with mixed conformations and sequential conformational changes. *Biochemistry* **23**, 2180–6.

Parish, N., Wrathmell, A., Hart, S. and Harris, J. E. (1986). The leucocytes of the elasmobranch *Scyliorhinus canicula* L. – a morphological study. *J. Fish Biol.* **28**, 545–61.

Payan, P. and Girard, J. P. (1977). Adrenergic receptors regulating patterns of blood flow through the gills of trout. *Am. J. Physiol.* **232**, H18–23.

Pearse, A. G. E. (1969). The cytochemistry and ultrastructure of polypeptide hormone-producing cells of the APUD series and the embryologic, physiologic and pathologic implications of the concept. *J. Histochem. Cytochem.* **17**, 303–13.

Pennec, J. P., Wardle, C. S., Harper, A. A. and Macdonald, A. G. (1988). Effects of high hydrostatic pressure on the isolated hearts of shallow water and deep sea fish; results of *Challenger* cruise 6B/85. *Comp. Biochem. Physiol.* **89A**, 215–18.

Perrier, H., Delcroix, J. P., Perrier, C. and Gras, J. (1973). An attempt to classify the plasma proteins of the rainbow trout (*Salmo gairdneri* Richardson) using disc electrophoresis, gel filtration and salt solubility fractionation. *Comp. Biochem. Physiol.* **46B**, 475–82.

Perrier, H., Delcroix, J. P., Perrier, C. and Gras, J. (1974). Disc electrophoresis of plasma proteins of fish. Physical and chemical characters; localization of fibrinogen, transferrin, and ceruloplasmin in the plasma of the rainbow trout (*Salmo gairdneri* Richardson). *Comp. Biochem. Physiol.* **49B**, 679–85.

Perrier, H., Perrier, C., Delcroix, J. P., Bornet, H. and Gras, J. (1976). Study of the iodide binding protein of the plasma of the rainbow trout (*Salmo gairdneri* Richardson). *Comp. Biochem. Physiol.* **55A**, 165–7.

Perrier, H., Perrier, C., Peres, G. and Gras, J. (1977). The perchlorosoluble proteins of the serum of the rainbow trout (*Salmo gairdneri* Richardson): albumin like and hemoglobin binding fraction. *Comp. Biochem. Physiol.* **57B**, 325–7.

Perry, S. F. and Vermette, M. G. (1987). The effects of prolonged epinephrine infusion on the physiology of the rainbow trout, *Salmo gairdneri*. 1. Blood respiratory, acid–base and ionic states. *J. exp. Biol.* **128**, 235–53.

Perry, S. F. and Walsh, P. J. (1989). Metabolism of isolated fish gill cells; contribution of epithelial chloride cells. *J. exp. Biol.* **144**, 507–20.

Perry, S. F., Daxboeck, C. and Dobson, G. P. (1985). The effect of perfusion flow rate and adrenergic stimulation on oxygen transfer in the isolated, saline-perfused head of the rainbow trout, (*Salmo gairdneri*). *J. exp. Biol.* **116**, 251–69.

Pettersson, K. (1983). Adrenergic control of oxygen transfer in pefused gills of the cod, *Gadus morhua*. *J. exp. Biol.* **102**, 327–35.

Pettersson, K. and Johansen, K. (1982). Hypoxic vasoconstriction and the effects of adrenaline on gas exchange efficiency in fish gills. *J. exp. Biol.* **97**, 263–72.

Peyraud-Waitzenegger, M., Barthelemy, L. and Peyraud, C. (1980). Cardiovascular and ventilatory effects of catecholamines in unrestrained eels (*Anguilla anguilla* L.). A study of seasonal changes in reactivity. *J. comp. Physiol.* **138B**, 367–75.

Pickering, A. D. and Pottinger, T. G. (1987a). Lymhocytopenia and interrenal activity during sexual maturation in the brown trout, *Salmo trutta* L. *J. Fish Biol.* **30**, 41–50.

Pickering, A. D. and Pottinger, T. G. (1987b). Crowding causes prolonged leucopenia in salmonid fish, despite interrenal acclimation. *J. Fish Biol.* **30**, 701–12.

Pickering, A. D. and Pottinger, T. G. (1988). Lymphocytopenia and the overwinter survival of Atlantic salmon parr, *Salmo salar* L. *J. Fish Biol.* **32**, 689–97.

Pohla, H., Lametschwandtner, A. and Adam, H. (1987). Die Vaskularistion der Kiemen von *Myxine glutinosa* L. (Cyclostomata). *Zool. scr.* **6**, 331–41.

Poole, C. A. and Satchell, G. H. (1979). Nociceptors in the gills of the dogfish *Squalus acanthias*. *J. comp. Physiol. A* **130**, 1–7.

Poupa, O. and Lindström, L. (1983). Comparative and scaling aspects of heart and body weights with reference to blood supply of cardiac fibres. *Comp. Biochem. Physiol.* **76A**, 413–21.

Poupa, O., Ask, J. A. and Helle, K. B. (1984). Absence of calcium paradox in the cardiac ventricle of the Atlantic hagfish (*Myxine glutinosa*). *Comp. Biochem. Physiol.* **78A**, 181–3.

Poupa, O., Gesser, H., Jonsson, S. and Sullivan, L. (1974). Coronary-supplied compact shell of ventricular myocardium in salmonids: growth and enzyme patterns. *Comp. Biochem. Physiol.* **48A**, 85–95.

Poupa, O., Helle, K. B. and Lomsky, M. (1985). Calcium paradox from cyclostome to man: a comparative study. *Comp. Biochem. Physiol.* **81A**, 801–5.

Powers, D. A. (1980). Molecular ecology of teleost fish hemoglobins: strategies for adapting to changing environments. *Amer. Zool.* **20**, 139–62.

Priede, I. G. (1974). The effect of swimming activity and section of the vagus nerves on heart rate in rainbow trout. *J. exp. Biol.* **60**, 305–19.

Priede, I. G. (1975). The blood circulatory function of the dorsal aorta ligament in rainbow trout (*Salmo gairdneri*). *J. Zool., Lond.* **175**, 39–52.

Priede, I. G.. (1976). Functional morphology of the bulbus arteriosus of rainbow trout. (*Salmo gairdneri* Richardson). *J. Fish Biol.* **9**, 209–16.

Primmett, D. R. N., Randall, D. J., Mazeaud, M. and Boutilier, R. G. (1986). The role of catecholamines in erythrocyte pH regulation and oxygen transport in rainbow trout (*Salmo gairdneri*) during exercise. *J. exp. Biol.* **122**, 139–48.

Pulsford, A., Fänge, R. and Morrow, W. J. R. (1982). Cell types and interactions in the spleen of the dogfish *Scyliorhinus canicula* L: an electron microscopic study. *J. Fish Biol.* **21**, 649–62.

Randall, D. J. and Daxboeck, C. (1982). Cardiovascular changes in the rainbow trout (*Salmo gairdneri* Richardson) during exercise. *Can. J. Zool.* **60**, 1135–40.

Randall, D. J. and Jones, D. R. (1973). The effect of deafferentiation of the pseudobranch on the respiratory response to hypoxia in the trout (*Salmo gairdneri*). *Respir. Physiol.* **17**, 291–301.

Randall, D. J. and Shelton, G. (1965). The effects of changes in environmental gas concentrations on the breathing and heart rate of a teleost fish. *Comp. Biochem. Physiol.* **9**, 229–39.

Randall, D. J. and Smith, J. C. (1967). The regulation of cardiac activity in fish in a hypoxic environment. *Physiol. Zool.* **40**, 104–13.

Randall, D. J. and Stevens, E. D. (1967). The role of adrenergic receptors in cardiovascular changes associated with exercise in salmon. *Comp. Biochem. Physiol.* **21**, 415–24.

Rang, H. P. and Dale, M. M. (1987). Chemical transmission and the autonomic nervous system. In *Pharmacology*, pp. 101–76. Churchill Livingston, Edinburgh and Melbourne.

Rasio, E. A., Bendayan, M. and Goresky, C. A. (1977). Diffusion permeability of an isolated rete mirabile. *Circulation Res.* **41**, 791–8.

Rasio, E. A., Bendayan, M. and Goresky, C. A. (1981). The effect of hyperosmolality on the permeability and structure of the capillaries of the isolated rete mirabile of the eel. *Circulation Res.* **49**, 661–76.

Rhodin, J. A. G. and Silversmith, C. (1972). Fine structure of elasmobranch arteries, capillaries and veins in the spiny dogfish, *Squalus acanthias*. *Comp. Biochem. Physiol.* **42A**, 59–64.

Ripplinger, J. and Pierron, J. C. (1973). Etude électrophysiologique, *in vitro*, de l'automatisme et des voies de conduction du ventricule de tanche (*Tinca tinca*). *C. R. Séances Soc. Biol.* **167**, 937–41.

Ristori, M. T. (1970). Réflexe de barosensibilité chez un poisson téléostéen (*Cyprinus carpio*). *C. R. Séances Soc. Biol.* **164**, 1512–16.

Ristori, M. and Dessaux, G. (1970). Sur l'existence d'un gradient de sensibilité dans les récepteurs branchiaux de *Cyprinus carpio*. *C. R. Séances Soc. Biol.* **164**, 1517–19.

Ristori, M. T. and Laurent, P. (1977). Action de l'hypoxie sur le système vasculaire branchial de la tête perfusée de truite. *C. R. Séances Soc. Biol.* **171**, 809–13.

Rombout, J. H. W. M. and van den Berg, A. A. (1989). Immunological importance of the second gut segment of carp. 1. Uptake and processing of antigens by epithelial cells and macrophages. *J. Fish Biol.* **35**, 13–22.

Root, R. W. (1931). The respiratory function of the blood of marine fishes. *Biol. Bull. mar. biol. Lab., Woods Hole* **61**, 427–56.

Rosenshein, I. L., Schluter, S. F. and Marchalonis, J. J. (1986). Conservation among the immunoglobulins of carcharine sharks and phylogenetic conservation of variable region determinants. *Vet. Immunol. Immunopathol.* **12**, 13–20.

Ross, L. G. (1979a). The haemodynamics of gas resorption from the physoclist swimbladder: the structure and morphometrics of the oval in *Pollachius virens* (L.). *J. Fish Biol.* **14**, 261–6.

Ross, L. G. (1979b). The haemodynamics of gas resorption from the physoclist swimbladder. II. The determination of blood flow rate in the oval of *Pollachius virens* (L.). *J. Fish Biol.* **14**, 389–93.

Ruud, J. T. (1954). Vertebrates without erythrocytes and blood pigment. *Nature, Lond.* **173**, 848–50.

Saetersdal, T. S., Justesen, N. P.. and Krohnstad, A. W. (1974). Ultrastructure and innervation of the teleostean atrium. *J. mol. Cell. Cardiol.* **6**, 415–37.

Sailendri, K. and Muthukkaruppan, V. (1975). Morphology of lymphoid organs in a cichlid teleost *Tilapia mossambica* (Peters). *J. Morph.* **147**, 109–22.

Saito, T. (1973). Effects of vagal stimulation on the pacemaker action potentials of carp heart. *Comp. Biochem. Physiol.* **44A**, 191–9.

Sandnes, K., Lie, Ø. and Waagbø, R. (1988). Normal ranges of some blood chemistry parameters in adult farmed Atlantic salmon, *Salmo salar. J. Fish Biol.* **32**, 129–36.

Santer, R. M. (1972). Ultrastructural and histochemical studies on the innervation of the heart of a teleost *Pleuronectes platessa* L. *Z. Zellforsch.* **131**, 519–28.

Santer, R. M. (1985). Morphology and innervation of the fish heart. *Advances in anatomy, embryology and cell biology* **89**, 1–102.

Santer, R. M. and Cobb, J. L. S. (1972). The fine structure of the heart of the teleost, *Pleuronectes platessa* L. *Z. Zellforsch.* **131**, 1–14.

Santer, R. M. and Greer Walker, M. (1980). Morphological studies on the ventricle of teleost and elasmobranch hearts. *J. Zool., Lond.* **190**, 259–72.

Santer, R. M., Greer Walker, M., Emerson, L. and Witthames, P. R. (1983). On the morphology of the heart ventricle in marine teleost fish (Teleostei). *Comp. Biochem. Physiol.* **76A**, 453–7.

Satchell, G. H. (1960). The reflex co-ordination of heart beat with respiration in the dogfish. *J. exp. Biol.* **37**, 719–31.

Satchell, G. H. (1961). The response of the dogfish to anoxia. *J. exp. Biol.* **38**, 531–43.

Satchell, G. H. (1962). Intrinsic vasomotion in the dogfish gill. *J. exp. Biol.* **39**, 503–12.

Satchell, G. H. (1965). Blood flow through the caudal vein of elasmobranch fish. *Aust. J. Sci.* **27**, 240–1.

Satchell, G. H. (1968). The genesis of certain cardiac arrhythmias in fish. *J. exp. Biol.* **49**, 129–41.

Satchell, G. H. (1970). A functional appraisal of the fish heart. *Fed. Proc.* **29**, 1120–3.

Satchell, G. H. (1971). *Circulation in Fishes.* Cambridge University Press, Cambridge.

Satchell, G. H. (1978). Type J receptors in the gills of fish. In *Studies in Neurophysiology* (presented to A. K. McIntyre), ed. R. Porter, pp.31–42. Cambridge University Press, Cambridge.

Satchell, G. H. (1984a). On the caudal heart of *Myxine* (Myxinoidea: Cyclostomata). *Acta zool., Stockh.* **65**, 125–33.

Satchell, G. H. (1984b). Respiratory toxicology of fishes. In *Aquatic Toxicology, Vol. 2*, ed. L. J. Weber, pp. 1–50. Raven Press, New York.

Satchell, G. H. (1986). Cardiac function in the hagfish, *Myxine*, (Myxinoidea: Cyclostomata). *Acta zool., Stockh.* **67**, 115–22.

Satchell, G. H. and Jones, M. P. (1967). The function of the conus arteriosus in the Port Jackson shark, *Heterodontus portusjacksoni. J. exp. Biol.* **46**, 373–82.

Satchell, G. H. and Weber, L. J. (1987). The caudal heart of the carpet shark, Cephaloscyllium isabella. *Physiol. Zool.* **60**, 692–8.

Sauer, J. and Harrington, J. P. (1988). Hemoglobins of the sockeye salmon, *Oncorhynchus nerka. Comp. Biochem. Physiol.* **91A**, 109–14.

Saunders, R. L. and Farrell, A. P. (1988). Coronary atherosclerosis in Atlantic salmon. No regression of lesions after spawning. *Atherosclerosis* **8**, 378–84.

Scheide, J. I. and Zadunaisky, J. A. (1988). Effect of atriopeptin II on the isolated opercular epitheium of *Fundulus heterocitus. Am. J. Physiol.* **254**, R27–32.

Schloesing, T. and Richard, J. (1898). Recherche de l'argon dans le gaz de la vessie natatoire des poissone et des physalies. *Compt. Rendu.* **122**, 615–17.

Scholander, P. F. (1957). The wonderful net. *Scientific American* **196**, 96–110.

Scholander, P. F. and Van Dam, L. (1954). Secretion of gases against high pressures in the swimbladder of deep sea fishes. II. The rete mirabile. *Biol. Bull. mar. biol. Lab., Wood Hole* **107**, 260–77.

Scott-Linthicum, D. and Carey, F. G. (1972). Regulation of brain and eye temperatures by the bluefin tuna. *Comp. Biochem. Physiol.* **43A**, 425–33.

Seibert, H. (1979). Thermal adaptation of heart rate and its parasympathetic control in the European eel, *Anguilla anguilla* (L.). *Comp. Biochem. Physiol.* **64C**, 275–8.

Seyama, I. and Irisawa, H. (1967). The effect of high sodium concentration on the action potential of the skate heart. *J. gen. Physiol.* **50**, 505–17.

Shabetai, R., Abdel, D. C. Graham, J. B., Bhargava, V., Keyes, R. S. and

Witztum, K. (1985). Function of the pericardium and pericardioperitoneal canal in elasmobranch fishes. *Am. J. Physiol.* **248**, H198–207.

Sharp, G. D. (1973). An electrophoretic study of hemoglobins of some scombroid fishes and related forms. *Comp. Biochem. Physiol.* **44B**, 381–8.

Shasby, D. M. and Roberts, R. L. (1987). Transendothelial transfer of macromolecules *in vitro*. *Fed. Proc.* **46**, 2506–10.

Shibata, Y. and Yamamoto, T. (1977). Gap junctions in the cardiac muscle cells of the lamprey. *Cell Tissue Res.* **178**, 477–82.

Short, S., Butler, P. J. and Taylor, E. W. (1977). The relative importance of nervous, humoral and intrinsic mechanisms in the regulation of heart rate and stroke volume in the dogfish *Scyliorhinus canicula*. *J. exp. Biol.* **70**, 77–92.

Skepper, J. N., Woodward, J. M. and Navaratnam, V. (1988). Immunocytochemical localization of natriuretic peptide sequences in the human right auricle. *J. mol. Cell. Cardiol.* **20**, 343–51.

Smeda, J. S. and Houston, A. H. (1979). Carbonic anhydrase (acetazolamide-sensitive esterase) activity in the red blood cells of thermally-acclimated rainbow trout, *Salmo gairdneri*. *Comp. Biochem. Physiol.* **62A**, 719–23.

Smirnov, V. N., Antonov, A. S., Antonova, G. N., Romanov, Y. A., Kabaeva, N. V., Tchertikhina, I. V. and Lukashev, M. E. (1989). Effects of forskolin and phorbol-myristate-acetate on cytoskeleton, extracellular matrix and protein phosphorylation in human endothelial cells. *J. mol. Cell Cardiol.* **21**, 3–11.

Smit, G. L. and Schoonbee, H. J. (1988). Blood coagulation factors in the freshwater fish *Oreochromis mossambicus*. *J. Fish Biol.* **32**, 673–7.

Smith, D. G. (1977). Sites of cholinergic vasoconstriction in trout gills. *Am. J. Physiol.* **233**, R222–9.

Smith, D. G. (1978). Neural regulation of blood pressure in rainbow trout. (*Salmo gairdneri*). *Can. J. Zool.* **56**, 1678–83.

Smith, D. G. and Chamley-Campbell, J. (1981). Localization of smooth-muscle myosin in branchial pillar cells of snapper (*Chrysophrys auratus*) by immunofluorescence histochemistry. *J. exp. Zool.* **215**, 121–4.

Smith, D. G., Nilsson, S., Wahlqvist, I. and Eriksson, B. M. (1985). Nervous control of the blood pressure in the Atlantic cod, *Gadus morhua*. *J. exp. Biol.* **117**, 335–47.

Smith, F. M. and Davie, P. S. (1984). Effects of sectioning cranial nerves IX and X on the cardiac response to hypoxia in the coho salmon, *Oncorhynchus kisutch*. *Can. J. Zool.* **62**, 766–8.

Smith, F. M. and Jones, D. R. (1978). Localization of receptors causing hypoxic bradycardia in trout (*Salmo gairdneri*). *Can. J. Zool.* **56**, 1260–5.

Smith, H. W. (1929). The composition of the body fluids of elasmobranchs. *J. biol. Chem.* **81**, 407–19.

Smith, M. A. K., McKay, M. C. and Lee, R. F. (1988). Catfish plasma lipoproteins: *in vivo* studies of apoprotein synthesis and catabolism. *J. exp. Zool.* **246**, 223–35.

Soivio, A. and Tuurala, H. (1981). Structural and circulatory responses to hypoxia in the secondary lamella of *Salmo gairdneri* gills at two temperatures. *J. comp. Physiol. B* **145**, 37–43.

Soivio, A., Nikinmaa, M. and Westman, K. (1980). The blood oxygen binding properties of hypoxic *Salmo gairdneri. J. comp. Physiol. B* **136**, 83–7.

Soivio, A., Westman, K. and Nyholm, K. (1974). The influence of changes in oxygen tension on the haematocrit value of blood samples from asphyxic rainbow trout (*Salmo gairdneri*). *Aquaculture* **3**, 395–401.

Stahl, B. J. (1974). *Vertebrate History: Problems in Evolution.* McGraw Hill, New York and London.

Starling, E. H. (1918). *The Law of the Heart.* Linacre lecture. Longmans Green, London.

Steen,J. B. (1963). The physiology of the swimbladder in the eel *Anguilla vulgaris.* III. The mechanism of gas secretion. *Acta physiol. scand.* **59**, 221–4.

Steen, J. B. (1970). The swimbladder as a hydrostatic organ. In *Fish physiology, Vol IV*, ed. W. S. Hoar and D. J. Randall, pp. 413–43. Academic Press, London and New York.

Steen, J. B. and Sund, T. (1977). Gas deposition by counter-current multiplication in the eel swim-bladder: experimental verification of a mathematical model. *J. Physiol., Lond.* **267**, 607–702.

Steffensen, J. F., Johansen, K., Sindberg, C. D., Sørensen, J. H. and Møller, J. L. (1984). Ventilation and oxygen consumption in the hagfish, *Myxine glutinosa* L. *J. exp. mar. Biol. Ecol.* **84**, 173–8.

Steffenssen, J. F., Lomholt, J. P. and Vogel, W. O. P. (1986). *In vivo* observations on a specialized microvasculature, the primary and secondary vessels in fishes. *Acta zool., Stockh.* **67**, 193–200.

Stensiö, E. A. (1932). *The Cephalaspids of Great Britain.* British Museum and B. Quaritch Ltd, London.

Stevens, E. D. (1973). The evolution of endothermy. *J. theor. Biol.* **38**, 597–611.

Stevens, E. D. and Carey, F. G. (1981). One why of the warmth of warm-bodied fish. *Am. J. Physiol.* **240**, R151–5.

Stevens, E. D. and Fry, F. E. J. (1971). Brain and muscle temperature in ocean-caught and captive skipjack tuna. *Comp. Biochem. Physiol.* **38A**, 203–11.

Stevens, E. D. and McLeese, J. M. (1984). Why bluefin tuna have warm tummies: temperature effect on trypsin and chymotrypsin. *Am. J. Physiol.* **246**, R487–94.

Stevens, E. D. and Neill, W. H. (1978). Body temperature relations of tunas, especially skipjack. In *Fish physiology, Vol VII*, ed. W. S. Hoar and D. J. Randall, pp. 315–59. Academic Press, New York and London.

Stevens, E. D. and Randall, D. J. (1967). Changes of gas concentrations in blood and water during moderate swimming activity in rainbow trout. *J. exp. Biol.* **46**, 329–37.

Stevens, E. D. and Sutterlin, A. M. (1976). Heat transfer between fish and ambient water. *J. exp. Biol.* **65**, 131–45.

Stevens, E. D., Bennion, G. R., Randall, D. J. and Shelton, G. (1972). Factors affecting arterial pressures and blood flow from the heart in intact unrestrained lingcod, *Ophiodon elongatus. Comp. Biochem. Physiol.* **43A**, 681–95.

Stevens, E. D., Lam, H. M. and Kendall, J. (1974). Vascular anatomy of the counter-current heat exchanger of skipjack tuna. *J. exp. Biol.* **61**, 145–53.

Stewart, J. M. and Driedzic, W. R. (1988). Fatty acid binding proteins in teleost fish. *Can. J. Zool.* **66**, 2671–5.

Stray-Pedersen, S. (1970). Vascular responses induced by drugs and by vagal stimulation in the swimbladder of the eel, *Anguilla vulgaris. Comp. gen. Pharmacol.* **1**, 358–64.

Stray-Pedersen, S. (1975). The effect of Ca^{++}, Mg^{++}, and H^{+} on the capillary permeability of the *rete mirabile* of the eel, *Anguilla vulgaris* L. *Acta physiol. scand.* **94**, 423–41.

Stray-Pedersen, S. and Nicolaysen, A. (1975). Qualitative and quantitative studies of the capillary structure in the *rete mirabile* of the eel, *Anguilla vulgaris* L. *Acta physiol. scand.* **94**, 339–57.

Stray-Pedersen, S. and Steen, J. B. (1975).The capillary permeability of the *rete mirabile* of the eel, *Anguilla vulgaris* L. *Acta physiol. scand.* **94**, 401–22.

Stuart, R. E., Hedtke, J. H. and Weber, L. J. (1983). Physiological and pharmacological investigation of the nonvascularized marine teleost heart with adrenergic and cholinergic agents. *Can. J. Zool.* **61**, 1944–8.

Suchard, M. E. (1907). Sur les valvules des veins de la grenouille. *C. R. Soc. Biol., Paris* **62**, 452–3.

Sulya, L. L., Box, B. E. and Gunter, G. (1961). Plasma proteins in the blood of fishes from the Gulf of Mexico. *Amer. J. Physiol.* **200**, 152–4.

Sund, T. (1977). A mathematical model for counter-current multiplication in the swim bladder. *J. Physiol., Lond.* **267**, 679–96.

Suzuki, K. (1986). Morphological and phagocytic characteristics of peritoneal exudate cells in tilapia, *Oreochromis niloticus* (Trewavas), and carp, *Cyprinus carpio. J. Fish Biol.* **29**, 349–64.

Tang, Y. and Boutilier, G. (1988). Correlation between catecholamine release and degree of acidotic stress in trout. *Am. J. Physiol.* **255**, R395–9.

Tang, Y., Nolan, S. and Boutilier, R. G. (1988). Acid-base regulation following acute acidosis in seawater-adapted rainbow trout, *Salmo gairdneri*: a possible role for catecholamines. *J. exp. Biol.* **134**, 297–312.

Tatner, M. F., Adams, A. and Leschen, W. (1987). An analysis of the primary and secondary antibody response in intact and thymectomized rainbow trout, *Salmo gairdneri* Richardson, to human gamma globulin and *Aeromonas salmonicida. J. Fish Biol.* **31**, 177–95.

Tebecis, A. K. (1967). A study of electrograms recorded from the conus arteriosus of an elasmobranch heart. *Aust. J. Biol. Sci.* **20**, 843–6.

Temma, K., Kishi, H., Kitizawz, T., Kondo, H., Ohta, T. and Katano, Y. (1986). Beta 2 adrenergic receptors are not only for circulating catecholamines in ventricular muscles of carp heart (*Cyprinus carpio*). *Comp. Biochem. Physiol.* **83C**, 265–9.

Tetens, V. and Christensen, N. J. (1987). Beta-adrenergic control of blood oxygen affinity in acutely hypoxia exposed rainbow trout. *J. comp. Physiol. B* **157**, 667–75.

Tetens, V. and Lykkeboe, G. (1981). Blood respiratory properties of rainbow trout, *Salmo gairdneri*: responses to hypoxia acclimation and anoxic incubation of blood in vitro. *J. comp. Physiol. B* **145**, 117–25.

Tetens, V. and Wells, R. M. G. (1984). Antarctic fish blood: respiratory properties and the effects of thermal acclimation. *J. exp. Biol.* **109**, 265–79.

Tetens, V., Lykkeboe, G. and Christensen, N. J. (1988). Potency of adrenaline and noradrenaline for β-adrenergic proton extrusion from red cells of rainbow trout, *Salmo gairdneri. J. exp. Biol.* **134**, 267–80.

Thomas, S., Fievet, B., Claireaux, G. and Motais, R. (1988). Adaptive respiratory responses of trout to acute hypoxia. 1. Effects of water ionic composition on blood acid-base response and gill morphology. *Respir. Physiol.* **74**, 77–90.

Thompson, D. W. (1917). *On Growth and Form.* Cambridge University Press, Cambridge.

Thorpe, A. and Ince, B. W. (1974). The effects of pancreatic hormones, catecholamines and glucose loading on blood metabolites in the northern pike (*Esox lucius*). *Gen. comp. Encodrinol.* **23**, 29–44.

Tota, B. (1983). Vascular and metabolic zonation in the ventricular myocardium of mammals and fishes. *Comp. Biochem. Physiol.* **76A**, 423–37.

Totland, G. K., Kryvi, H., Bone, Q. and Flood, P. R. (1981). Vascularization of the lateral muscles of some elasmobranchiomorph fishes. *J. Fish Biol.* **18**, 223–34.

Tufts, B. L. and Randall, D. J. (1989). The functional significance of adrenergic pH regulation in fish erythrocytes. *Can. J. Zool.* **67**, 235–8.

Tysekiewicz, K. (1969). Structure and vascularisation of the skin of the pike (*Esox lucius* L.). *Acta biol. cracoviensia., Zoologia* **12**, 67–79.

van Dijk, P. L. M. and Wood, C. M. (1988). The effect of β-adrenergic blockade on the recovery process after strenuous exercise in the rainbow trout, *Salmo gairdneri* Richardson. *J. Fish Biol.* **32**, 557–70.

Vassalle, M. (1977). The relationship among cardiac pacemakers. *Circ. Res.* **41**, 269–77.

Vecchio, P. J. Del., Siflinger-Birnboim, A., Shepard, J. M., Bizois, R., Cooper, J. A. and Malik, A. B. (1987). Endothelial monolayer permeability to macromolecules. *Fed. Proc.* **46**, 2511–15.

Vogel, W. O. P. (1978). The origin of Fromm's arteries in trout gills. *Z. Mikrosk. Arch. Forsk. Leipzig* **92**, 566–70.

Vogel, W. O. P. (1981). Struktur und organisationsprinzip im Gefasssystem der Knochenfische. *Gegeanbaurs morph. Jahrb., Leipzig* **127**, 772–84.

Vogel, W. O. P. (1985a). Systemic vascular anastomoses, primary and secondary vessels in fish, and the phylogeny of lymphatics. In *Cardiovascular Shunts, Alfred Benzon Symp. 21*, ed. K. Johansen and W. W. Burggren, pp. 143–59. Munksgaard, Copenhagen.

Vogel, W. O. P. (1985b). The caudal heart of fish: not a lymph heart. *Acta anat.* **121**, 41–5.

Vogel, W. O. P. and Claviez, M. (1981). Vascular specialization in fish but no evidence for lymphatics. *Z. Naturforsch.* **36**, 490–2.

Vogel, W., Vogel, V. and Schlote, W. (1974). Ultrastructural study of arteriovenous anastomoses in gill filaments of *Tilapia mossambica. Cell Tissue Res.* **155**, 491–512.

Wahlqvist, I. (1980). Effects of catecholamines on isolated systemic and branchial vascular beds of the cod, *Gadus morhua. J. comp. Physiol. B* **137**, 139–43.

Wahlqvist, I. (1981). Branchial vascular effects of catecholamines released from the head kidney of the atlantic cod *Gadus morhua. Mol. Physiol.* **1**, 235–41.

Wahlqvist, I. and Nilsson, S. (1977). The role of sympathetic fibres and circulating catecholamines in controlling the blood pressure and heart rate in the cod, *Gadus morhua. Comp. Biochem. Physiol.* **57C**, 65–7.

Wahlqvist, I. and Nilsson, S. (1980). Adrenergic control of the cardio-vascular system of the Atlantic cod, *Gadus morhua*, during 'stress'. *J. comp. Physiol.* **137**, 145–50.

Wahlqvist, I. and Nilsson, S. (1981). Sympathetic nervous control of the vasculature in the tail of the Atlantic cod, *Gadus morhua. J. comp. Physiol. B* **144**, 153–6.

Watson, A. D. and Cobb, J. L. S. (1979). A comparative study on the innervation and the vascularization of the bulbus arteriosus in teleost fish. *Cell Tissue Res.* **196**, 337–46.

Weber, R. E. and Lykkeboe, G. (1978). Respiratory adaptations in carp blood. Influence of hypoxia, red cell organic phosphates, divalent cations and CO_2 on hemoglobin oxygen affinity. *J. comp. Physiol. B* **128**, 127–37.

Weber, R. E., Johansen, K., Lykkeboe, G. and Maloiy, G. M. O. (1977). Oxygen-binding properties of hemoglobins from estivating and active African lungfish. *J. exp. Zool.* **199**, 85–96.

Weber, R. E., Wells, R. M. G. and Rosetti, J. E. (1983). Allosteric interactions governing oxygen equilibria in the haemoglobin system of the spiny dogfish, *Squalus acanthias. J. exp. Biol.* **103**, 109–20.

Weber, R. E., Wood, S. C. and Lomholt, J. P. (1976). Temperature acclimation and oxygen-binding properties of blood and multiple haemoglobins of rainbow trout. *J. exp. Biol.* **65**, 333–45.

Weeks, A., Warrinner, J. E., Mason, P. L. and McGinnis, D. S. (1986). Influence of toxic chemicals on the chemotactic response of fish macrophages. *J. Fish Biol.* **28**, 653–8.

Weinberg, S. R., Siegel, C. D. and Gordon, A. S. (1973). Studies on the peripheral blood cell parameters and morphology of the red paradise fish *Macropodus opercularis*. Effect of food deprivation on erythropoiesis. *Anat. Rec.* **175**, 7–13.

Wells, R. M. G. (1989). Hemoglobin physiology in vertebrate animals: a cautionary approach to adaptionist thinking. In *Advances in Environmental and Comparative Physiology. Vertebrate gas exchange: from environment to cell*, ed. R. G. Boutilier, pp. 1–37. Springer-Verlag, Berlin.

Wells, R. M. G. and Forster, M. E. (1989). Dependence of blood viscosity on haematocrit and shear rate in a primitive vertebrate. *J. exp. Biol.* **145**, 483–7.

Wells, R. M. G. and Weber, R. E. (1983). Oxygenational properties and phosphorylated metabolic intermediates in blood erythrocytes of the dogfish, *Squalus acanthias. J. exp. Biol.* **103**, 95–108.

Wells, R. M. G., Ashby, M. D., Duncan, S. J. and Macdonald, J. A. (1980). Comparative study of the erythrocytes and haemoglobins in nototheniid fishes from Antarctica. *J. Fish Biol.* **17**, 517–27.

Wells, R. M. G., Forster, M. E., Davison, W., Taylor, H. H., Davie, P. S. and Satchell, G. H. (1986). Blood oxygen transport in the free-swimming hagfish, *Eptatretus cirrhatus. J. exp. Biol.* **123**, 45–53.

Wells, R. M. G., Grigg, G. C., Beard, L. A. and Summers, G. (1989). Hypoxic responses in a fish from a stable environment: blood oxygen transport in the Antarctic fish *Pagothenia borchgrevinki. J. exp. Biol.* **141**, 97–111.

Wittenberg, J. B. and Wittenberg, B. A. (1962). Active secretion of oxygen into the eye of fish. *Nature, Lond.* **194**, 106–7.

Wittenberg, J. B. Schwend, M. J. and Wittenberg, B. A. (1964). The secretion of oxygen into the swimbladder of fish. III. The role of carbon dioxide. *J. gen. Physiol.* **48**, 337–55.

Wittenberger, C., Coprean, D. and Morar, L. (1975). Studies on the carbohydrate metabolism of the lateral muscles of carp (influence of phloridzin, insulin and adrenaline). *J. Comp. Physiol.* **101**, 161–72.

Wood, C. M. (1974a). A critical examination of the physical and adrenergic factors affecting blood flow through the gills of rainbow trout. *J. exp. Biol.* **60**, 241–65.

Wood, C. M. (1974b). Mayer waves in the circulation of a teleost fish. *J. exp. Zool.* **189**, 267–74.

Wood, C. M. (1975). A pharmacological analysis of the adrenergic and cholinergic mechanisms regulating branchial vascular resistance in the rainbow trout (*Salmo gairdneri*). *Can. J. Zool.* **53**, 1569–77.

Wood, C. M. and Shelton, G. (1975). Physical and adrenergic factors affecting systemic vascular resistance in the rainbow trout: a comparison with branchial vascular resistance. *J. exp. Biol.* **63**, 505–23.

Wood, C. M. and Shelton, G. (1980a). The reflex control of heart rate and cardiac output in the rainbow trout: interactive influences of hypoxia, haemorrhage and systemic vasomotor tone. *J. exp. Biol.* **87**, 271–84.

Wood, C. M. and Shelton, G. (1980b). Cardiovascular dynamics and adrenergic responses of the rainbow trout *in vivo*. *J. exp. Biol.* **87**, 247–70.

Wood, C. M., McMahon, B. R. and McDonald, D. G. (1979). Respiratory, ventilatory and cardiovascular responses to experimental anaemia in the starry flounder, *Platichthys stellatus*. *J. exp. Biol.* **82**, 139–62.

Yamamoto, K. (1988). Contraction of spleen in exercised freshwater teleost. *Comp. Biochem. Physiol.* **89A**, 65–6.

Yamamoto, K., Itazawa, Y. and Kobayashi, H. (1985). Direct observation of fish spleen by an abdominal window method and its application to exercised and hypoxic yellow tail. *Jap. J. Ichthyol.* **31**, 427–33.

Yamamoto, K. and Itazawa, Y. (1989). Erythrocyte supply from the spleen of exercised carp. *Comp. Biochem. Physiol.* **92A**, 139–44.

Yamauchi, A. (1980). Fine structure of the fish heart. In *Hearts and Heart-Like Organs, Vol. 1*, ed. G. H. Bourne, pp. 119–48. Academic Press, New York and London.

Yamauchi, A. and Burnstock, G. (1968). An electron microscopic study on the innervation of the trout heart. *J. comp. Neurol.* **132**, 567–88.

Yamauchi, A., Fujimaki, Y. and Yokoto, R. (1973). Fine structural studies of the sino-auricular nodal tissue in the heart of a teleost fish, *Misgurnus*, with particular reference to the cardiac internuncial cell. *Am. J. Anat.* **138**, 407–30.

Zapata, A. (1979a). Ultrastructural study of the teleost fish kidney. *Dev. comp. Immunol.* **3**, 55–65.

Zapata, A. (1979b). Ultrastucture of the gut associated lymphoid tissue (GALT) of *Rutilus rutilus*. *Morf. normal Path. Soc. A* **3**, 23–9.

Zuberbuhler, R. C. and Bohr, D. F. (1965). Responses of coronary smooth muscle to catecholamines. *Circulation Res.* **16**, 431–9.

APPENDIX OF POPULAR AND SCIENTIFIC NAMES

Popular name	Scientific name	Page
African lungfish	*Protopterus ethiopicus*	16, 18, 63
Albacore	*Thunnus alulunga*	12, 13, 24, 67, 69, 123
Angel shark	*Squatina squatina*	86
Angler fish	*Lophius piscatorius*	28, 61, 82, 103
Arctic char	*Salvelinus alpinus*	76, 194
Arctic grayling	*Thymallus arcticus*	104
Argentine	*Argentina silus*	34
Asiatic eel	*Anguilla japonica*	31, 32
Atlantic hagfish	*Myxine glutinosa*	11, 12, 14, 18, 20, 24, 35, 37, 38, 45, 53, 59, 83, 91, 129, 136, 137, 138, 139, 140, 144, 150, 159, 160, 183–197
Atlantic salmon	*Salmo salar*	33, 34, 36, 60, 72, 147, 158, 159
Atlantic tarpon	*Megalops atlanticus*	31
Australian eel	*Anguilla australis*	37, 38, 78, 82
Barn door skate	*Raja stabuliforis*	27
Benthic fish	*Antimora rostrata*	110
Bichir	*Polypterus bichir*	31, 86
Big-eye tuna	*Thunnus obesus*	118, 124
Big skate	*Raja binoculata*	178
Black bullhead	*Ictalurus melas*	95, 98
Black mouthed dogfish	*Galleus melanostomus*	105
Black prickleback	*Xiphister atropurpureus*	114
Black skipjack tuna	*Euthynnus alleteratus*	124
Bluefin tuna	*Thunnus thynnus*	118, 119, 120, 121, 124
Blue fish	*Pomatomus saltatrix*	5, 75, 98, 112, 113, 124
Blue marlin	*Makaira nigricans*	31, 60, 71, 124, 189
Blue shark	*Prionace glauca*	86
Blue skate	*Raja batis*	87
Bowfin	*Amia calva*	96, 113
Bream	*Chrysophys auratus*	91
Brill	*Rhombus rhombus*	103
Brook trout	*Salvelinus fontinalis*	36

Brown bullhead	*Ictalurus nebulosus*	17, 18, 188
Brown trout	*Salmo trutta*	83
Buffalo sculpin	*Enophrys bison*	160
Burbot	*Lota lota*	64
Carp	*Cyprinus carpio*	16, 17, 18, 27, 42, 63, 64, 65, 66, 68, 84, 85, 101, 110, 115, 146, 167, 176, 180, 186
Carpet shark	*Cephaloscyllium isabella*	127, 133, 134, 135
Catfish	*Clarias gariepinus*	20
Cephalaspidomorph	*Hemicyclaspis murchisoni*	195
Channel catfish	*Ictalurus punctatus*	34, 72, 94, 176
Chimaera	*Chimaera monstrosa*	12, 26, 103
Chinook salmon	*Oncorhynchus tshawytscha*	33, 70, 160
Chum salmon	*Oncorhynchus keta*	33
Cichlid	*Oreochromis grahami*	1
Clear ray	*Raja diaphenes*	27
Cod	*Gadus morhua*	8, 29, 34, 36, 37, 48, 53, 93, 145, 147, 148, 161, 162, 163, 164, 168, 174, 177
Coho salmon	*Oncorhynchus kisutch*	69, 153
Common eel	*Anguilla anguilla*	12, 18, 24, 44, 49, 95, 101, 102, 107, 108, 110, 115, 117, 135, 136, 148, 151, 155, 166, 176, 191
Conger eel	*Conger conger*	33
Crucian carp	*Carassius carassius*	105
Cusk eel	*Ophidium barbatum*	102
Dover sole	*Microstomus pacificus*	65, 114
Flounder	*Platichthys flesus*	36
Frigate mackerel	*Auxis rochei*	60
Glass catfish	*Kryptopterus bicirrhis*	98
Glass whiting	*Haletta semifasciata*	104
Golden bream	*Chrysophrys auratus*	91
Goldfish	*Carassius auratus*	36, 65, 77, 84, 181
Goldsinny wrasse	*Ctenolabrus rupestris*	95, 148
Gourami	*Osphronemus goramy*	95
Greenfish	*Hexagrammus* sp.	101
Haddock	*Melanogrammus aeglefinus*	12
Hake	*Merluccius merluccius*	10
Herring	*Clupea harengus*	35, 43, 61
Horn shark	*Heterodontus francisci*	30, 60
Icefish	*Chaenocephalus aceratus*	60
Japanese lamprey	*Entosphenus japonicus*	14
Japanese loach	*Misgurnus anguillicaudatus*	16
Japanese skate	*Dasyatis akajei*	20
Kawakawa	*Euthynnus affinis*	37, 64, 69

Lamprey	*Petromyzon marinus*	43, 83, 184
Lancelet	*Branchiostoma lanceolatum*	31
Lingcod	*Ophiodon elongatus*	27, 37, 46, 53, 76, 91, 93, 100, 101, 161, 171, 173
Little skate	*Raja erinacea*	144, 180
Long-jawed goby	*Gillichthys mirabilis*	100
Long-nosed gar	*Lepisosteus osseus*	31
Long-nosed skate	*Raja nasuta*	146, 147
Mackerel	*Scomber scombrus*	9, 33, 36, 37, 60, 61, 98
Mako shark	*Isurus oxyrinchus*	116, 120, 122, 124
Medaka	*Oryzias latipes*	15, 27
Menhaden	*Brevoortia tyrannus*	65
Molly	*Poecilia latipinna*	100
Mullet	*Mugil cephalus*	188
New Zealand hagfish	*Eptatretus cirrhatus*	20, 31, 35, 53, 62, 64, 71, 83, 140, 184, 185, 188, 189, 194, 196
Nototheniids	*Dissostichus mawsoni*	60, 62, 65
	Pagothenia bernacchii	60
	Pagothenia borchgrevinki	64, 68
Nurse shark	*Ginglymostoma cirratum*	86
Ocean pout	*Macrozoaraces americanus*	12, 35, 40, 103, 166, 185
Opaleye	*Girella nigricans*	19
Pacific dogfish	*Squalus suckleyi*	178
Pacific hagfish	*Eptatretus stouti*	17, 40, 138, 186, 188, 189, 192, 193
Paddlefish	*Polyodon spathula*	85
Perch	*Perca perca*	12, 36, 43, 44, 107, 137, 155, 164
Pike	*Esox esox*	12, 43, 44, 77, 158, 159, 180
Pipe fish	*Syngnathus* sp.	103
Plaice	*Pleuronectes platessa*	12, 15, 17, 34, 36, 61, 80, 81, 85, 176
Pond loach	*Misgurnus fossilis*	103
Porbeagle shark	*Lamna nasus*	116, 120, 122, 124
Port Jackson shark	*Heterodontus portusjacksoni*	20, 29, 30, 128, 129, 132, 133, 134
Puffer	*Sphaeroides maculatus*	112
Red band fish	*Cepola rubescens*	102
Red paradise fish	*Macropodus opercularis*	60, 61, 188
Remora	*Echeneis naverales*	112, 113
Rough-tailed stingray	*Dasyatis centroura*	59, 188
Rudd	*Scardinus erythrophthalmus*	102
Saithe	*Pollachius virens*	148
Salt water minnow	*Fundulus heteroclitus*	60, 150
Scorpion fish	*Scorpaenichthys* sp.	101
Sea horse	*Hippocampus* sp.	103
Sea perch	*Sebastodes* sp.	101

Sea raven	*Hemitripterus americanus*	12, 24, 35, 37, 38, 39, 40, 116, 160, 175, 176, 185
Shad	*Alosa alosa*	43
Short-finned eel	*Anguilla australis schmidtii*	130, 167
Shorthorn sculpin	*Myoxocephalus scorpius*	76, 194
Skipjack tuna	*Katsuwonus pelamis*	91, 116, 118, 123
Smooth hound	*Mustelus antarcticus*	133
Smooth sculpin	*Artedius lateralis*	91
Snapper	*Lutianus rivularis*	80
Sockeye salmon	*Oncorhynchus nerka*	69
Sole	*Soles solea*	91
Spiny dogfish	*Squalus acanthias*	12, 20, 30, 33, 42, 44, 51, 59, 60, 63, 64, 69, 94, 127, 128, 151, 154, 155, 156, 175, 185
Spotted dogfish	*Scyliorhinus canicula*	12, 33, 37, 53, 77, 78, 79, 80, 82, 86, 87, 88, 153, 161, 163, 171, 175, 177, 180, 181
Starry dogfish	*Scyliorhinus stellaris*	26, 180
Starry flounder	*Platichthys stellatus*	34, 48, 126, 127
Starry ray	*Raja radiata*	144
Stickleback	*Gastrosteus aculeatus*	102
Stingrays	*Urolophus* spp.	86, 95
Striped bass	*Morone saxatilis*	115
Striped marlin	*Tetrapterus audax*	124
Sturgeon	*Acipenser stellatus*	78, 86, 87, 104, 105
Swordfish	*Xiphias gladius*	31, 33, 124, 125
Tench	*Tinca tinca*	18, 59, 60, 64, 66, 67, 101, 102, 137, 189
Thresher shark	*Alopias vulpinus*	33
Thornback ray	*Platyrhinoidis triseriata*	24
Tilapia	*Oreochromis mossambicus*	74, 78, 82, 93, 98, 100, 164
Toad fish	*Opsanus beta*	93
Torpedo	*Torpedo* sp.	33
Trout = Rainbow trout	*Salmo gairdneri*	11, 12, 14, 17, 18, 24, 27, 33, 34, 35, 36, 37, 40, 43, 53, 58, 63, 64, 66, 67, 68, 70, 72, 75, 81, 83, 89, 91, 92, 94, 95, 96, 98, 111, 113, 114, 115, 127, 141, 146, 147, 148, 150, 151, 152, 153, 155, 158, 159, 160, 161, 162, 163, 164, 165, 166, 167, 168, 171, 173, 174, 175, 176, 177, 178, 179, 180, 181, 188
Turbot	*Rhombus maximus*	36, 80, 103
Velvet belly shark	*Etmopterus spinax*	105
White fish	*Coreogonus* sp.	81
White pointer shark	*Carcharodon carcharias*	116, 120, 128
White sturgeon	*Acipenser transmontanus*	86, 87
Whiting	*Gadus merlangus*	6

Winter flounder	*Pseudopleuronectes*	48, 75, 76, 77, 114, 127,
	americanus	174, 185
Wrasses	*Labrus, Crenilabrus*	89, 111, 148
Xenacanthid shark	*Xenacanthus* sp.	129
Yellowfin croaker	*Umbrina roncador*	19, 186
Yellowfin tuna	*Thunnus albacares*	60
Yellow tail	*Seriola quinqueradiata*	68, 75, 167
Zebra fish	*Zebra danio*	12
Innominate species	*Gadus thori*	12

INDEX